American Society of Civil Engineers

Standard Practice for the Design and Operation of Hail Suppression Projects

This document uses both Système International (SI) units and customary units.

Published by the American Society of Civil Engineers
1801 Alexander Bell Drive
Reston, Virginia 20191-4400

Abstract

This document describes a process through which hail suppression operations should be designed, organized, and conducted. The information contained herein is intended to be helpful to those persons wishing to implement operational hail suppression activities, and provides information on the design, conduct, and evaluation of such efforts. While not a technical meteorological monograph on the subject, it is intended to provide the best scientific information currently available on the subject to optimize the likelihood of success.

Library of Congress Cataloging-in-Publication Data

Environmental and Water Resources Institute (U.S.). Hail Suppression Standards Subcommitee.

   Standard practice for the design and operation of hail suppression projects : a document for consideration by the ASCE Atmospheric Water Management Standards Committee / prepared by the Hail Suppression Standards Subcommittee.
      p. cm.—(ASCE standard)
   "EWRI/ASCE 39-03."
   Includes bibliographical references and index.
   ISBN 0-7844-0651-0
      1. Hail control. I. Environmental and Water Resources Institute (U.S.). Atmospheric Water Management Standards Committee. II. Title.

   QC929.H15E58 2003
   551.68'76—dc21

                                                        2002043889

STANDARDS

In April 1980, the Board of Direction approved ASCE Rules for Standards Committees to govern the writing and maintenance of standards developed by the Society. All such standards are developed by a consensus standards process managed by the Management Group F (MGF), Codes and Standards. The consensus process includes balloting by the balanced standards committee made up of Society members and nonmembers, balloting by the membership of ASCE as a whole, and balloting by the public. All standards are updated or reaffirmed by the same process at intervals not exceeding 5 years.

The following Standards have been issued:

ANSI/ASCE 1-82 N-725 Guideline for Design and Analysis of Nuclear Safety Related Earth Structures
ANSI/ASCE 2-91 Measurement of Oxygen Transfer in Clean Water
ANSI/ASCE 3-91 Standard for the Structural Design of Composite Slabs and ANSI/ASCE 9-91 Standard Practice for the Construction and Inspection of Composite Slabs
ASCE 4-98 Seismic Analysis of Safety-Related Nuclear Structures
Building Code Requirements for Masonry Structures (ACI 530-02/ASCE 5-02/TMS 402-02) and Specifications for Masonry Structures (ACI 530.1-02/ASCE 6-02/TMS 602-02)
SEI/ASCE 7-02 Minimum Design Loads for Buildings and Other Structures
ANSI/ASCE 8-90 Standard Specification for the Design of Cold-Formed Stainless Steel Structural Members
ANSI/ASCE 9-91 listed with ASCE 3-91
ASCE 10-97 Design of Latticed Steel Transmission Structures
SEI/ASCE 11-99 Guideline for Structural Condition Assessment of Existing Buildings
ANSI/ASCE 12-91 Guideline for the Design of Urban Subsurface Drainage
ASCE 13-93 Standard Guidelines for Installation of Urban Subsurface Drainage
ASCE 14-93 Standard Guidelines for Operation and Maintenance of Urban Subsurface Drainage
ASCE 15-98 Standard Practice for Direct Design of Buried Precast Concrete Pipe Using Standard Installations (SIDD)
ASCE 16-95 Standard for Load and Resistance Factor Design (LRFD) of Engineered Wood Construction

ASCE 17-96 Air-Supported Structures
ASCE 18-96 Standard Guidelines for In-Process Oxygen Transfer Testing
ASCE 19-96 Structural Applications of Steel Cables for Buildings
ASCE 20-96 Standard Guidelines for the Design and Installation of Pile Foundations
ASCE 21-96 Automated People Mover Standards—Part 1
ASCE 21-98 Automated People Mover Standards—Part 2
ASCE 21-00 Automated People Mover Standards—Part 3
SEI/ASCE 23-97 Specification for Structural Steel Beams with Web Openings
SEI/ASCE 24-98 Flood Resistant Design and Construction
ASCE 25-97 Earthquake-Actuated Automatic Gas Shut-Off Devices
ASCE 26-97 Standard Practice for Design of Buried Precast Concrete Box Sections
ASCE 27-00 Standard Practice for Direct Design of Precast Concrete Pipe for Jacking in Trenchless Construction
ASCE 28-00 Standard Practice for Direct Design of Precast Concrete Box Sections for Jacking in Trenchless Construction
SEI/ASCE/SFPE 29-99 Standard Calculation Methods for Structural Fire Protection
SEI/ASCE 30-00 Guideline for Condition Assessment of the Building Envelope
SEI/ASCE 32-01 Design and Construction of Frost-Protected Shallow Foundations
EWRI/ASCE 33-01 Comprehensive Transboundary International Water Quality Management Agreement
EWRI/ASCE 34-01 Standard Guidelines for Artificial Recharge of Ground Water
EWRI/ASCE 35-01 Guidelines for Quality Assurance of Installed Fine-Pore Aeration Equipment
CI/ASCE 36-01 Standard Construction Guidelines for Microtunneling
SEI/ASCE 37-02 Design Loads on Structures During Construction
CI/ASCE 38-02 Standard Guideline for the Collection and Depiction of Existing Subsurface Utility Data
EWRI/ASCE 39-03 Standard Practice for the Design and Operation of Hail Suppression Projects

FOREWORD

In April 1995, the Board of Direction approved the revision to the ASCE Rules for Standards Committees to govern the writing and maintenance of standards developed for the Society. All such standards are developed by a consensus standards process managed by the ASCE Codes and Standards Activities Committee (CSAC). The consensus process includes balloting by a balanced ASCE Environmental and Water Resources Institute (EWRI), Atmospheric Water Management (AWM) Standards Committee (SC), made up of Society members and nonmembers, balloting by the membership of ASCE as a whole, and balloting by the public. All standards are updated or reaffirmed by the same process at intervals not exceeding five years.

The provisions of this document have been written in permissive language and, as such, offer to the user a series of options or instructions but do not prescribe a specific course of action. Significant judgement has been left to the user of this document.

This document describes a process through which hail suppression operations should be designed, organized, and conducted. The information contained herein is intended to be helpful to those persons wishing to implement operational hail suppression activities, and provides information on the design, conduct, and evaluation of such efforts. While not a technical meteorological monograph on the subject, it is intended to provide the best scientific information currently available on the subject to optimize the likelihood of success. ASCE Manual No. 81, *Guidelines for Cloud Seeding to Augment Precipitation* (Kahan et al. 1995), is referenced in many sections of this guideline.

This standard has been prepared in accordance with the CSAC Style Manual, 4 September 1998 revision, with recognized engineering principles, and should not be used without the user's competent knowledge for a given application. The International System of Units (SI units) is used throughout, with English equivalents also provided. Exceptions are the use of the Celsius (°C) temperature scale and, where appropriate, centimeters (cm) in lieu of meters (m). Section 8 provides all conversion factors used herein. Italics denote special emphasis.

The publication of this standard by ASCE is not intended as warrant that the information contained therein is suitable for any general or specific use, and the Society takes no position respecting the validity of patent rights. The user is advised that the determination of patent rights or risk of infringement is entirely their own responsibility.

Many contributed materially to this document by their comments, review, illustrations, and photographs. The primary authors of this document were the ASCE EWRI AWM SC Hail Suppression Subcommittee members: Bruce A. Boe (chair), Arnett S. Dennis, Andrew G. Detwiler, Thomas J. Henderson, Terrence W. Krauss, Griffith M. Morgan, James H. Renick, and José Luís Sánchez. Others who contributed materially include Magomet T. Abshaev, Thomas P. DeFelice, Richard D. Farley, William G. Finnegan, Don A. Griffith, Conrad G. Keyes, Jr., Nancy C. Knight, Fred J. Kopp, Harold D. Orville, Petio Simeonov, Paul L. Smith, and Richard H. Stone.

CONTENTS

LIST OF FIGURES

LIST OF TABLES

Standard Practice for the Design and Operation of Hail Suppression Projects

1.0 THE HAIL PROBLEM

Unlike other forms of precipitation, hail comes in bursts of short duration, falls over isolated and unpredictable locations, and usually occupies only a small percentage of the total area covered by precipitation during any given storm. Nonetheless, the sudden, instantly visible destruction caused by hail can have devastating financial and emotional impacts. It is not surprising, therefore, that people in many places have attempted to avert hail damage to their crops and other property.

1.1 HISTORICAL PERSPECTIVE

People in ancient times employed magic, including incantations, shouted and shook their fists, hurled lances, and shot arrows skyward in attempts to rout oncoming hailstorms. In Europe in the Middle Ages, Christians resorted to prayer and the ringing of church bells. It is not clear whether the ringing of the bells was expected to produce a physical effect or was viewed simply as an invocation of God's protection against impending disaster.

The introduction of gunpowder led to another approach to hail suppression. It was thought by some persons that explosions could somehow disrupt the hail formation process, and so the hail cannon was developed. Hundreds of cannon were in operation, mainly in European countries near the Alps, in the years just before World War I, and their use continued on a smaller scale after that conflict ended. A number of theories were advanced to justify the use of hail cannons. For example, shock waves were postulated either to disrupt the updraft required for hail formation, to soften the falling hailstones, or to induce the freezing of supercooled water droplets essential for hail development. Whether or how these effects might be adequately achieved by cannons was never resolved.

Small rockets containing explosive charges and developed in Italy were an outgrowth of the hail cannon technology. As early as 1937, these devices were being fired at storms from the ground. France, Switzerland, Kenya, and South Africa followed with similar trials, some of which lasted until 1960 or later. Some users of the devices argued that the shock waves they produced could soften or even shatter hailstones, especially if the hailstones contained pockets of liquid water.

Modern hail suppression technology is based on discoveries at the General Electric Research Laboratories in the late 1940s (Schaefer 1946, Vonnegut 1947) regarding the ice nucleating properties of dry ice and silver iodide. As it was well known that hailstone growth depended upon a supply of supercooled water in the hail-producing clouds, it was reasoned that artificial freezing of that supercooled water would suppress the formation and growth of hail. Commercial hail suppression operations involving the release of silver iodide from ground-based generators and from aircraft became quite common in the United States during the 1950s. Experimental programs were undertaken in many other countries as well. A few such programs are mentioned below. The most extensive projects were undertaken by scientists and engineers in the U.S.S.R., who pioneered new techniques for delivery of silver iodide or other seeding agents to target clouds. By the 1960s, their arsenal of anti-hail weapons included artillery shells and quite sophisticated rockets. On the basis of comparisons of crop damage in protected areas and in neighboring unprotected areas, substantial hail suppression effects were claimed (Atlas 1965, Battan 1965).

An extensive network of ground-based silver iodide generators has been used to suppress hail in southwestern France since 1951. Although targeting is less certain than with rockets or aircraft delivery systems, the program's operators consider it successful. For the period from 1965 through 1982, they report a decrease in hail damage on the order of 40% (Dessens 1986), and more recently, for the years 1988–1995, a decrease of 42% (Dessens 1998).

One of the most extensive hail suppression research projects began in Alberta, Canada, in 1956. In one of the regions of greatest hail frequency in the world, the hail insurance premiums approached one-quarter of the crop value. Airborne seeding served as the major method of dispensing the silver iodide nuclei. Although a research project was conducted for more than 30 years, with 14 years of cloud seeding, the participating scientists were unable to reach a conclusion about the effectiveness of hail suppression. The Alberta research effort is currently inactive, but an operational hail suppression project was launched in 1996, funded entirely by the property and casualty insurance companies (Krauss and Renick 1997).

In Kenya, Africa, a 7-yr operational hail suppression program was conducted over tea estates where the hail frequency averages nearly 200 days per year. Pyrotechnics containing silver iodide were burned on aircraft flying near cloud base. Because tea is harvested roughly every 15 days, program evaluations included comparisons of crop production from areas under protection with areas not targeted by the seeding program. These simple evaluations suggested reductions in hail damage of 64%, but the program design did not allow for more sophisticated statistical tests.

A Swiss hail suppression experiment called Grossversuch III was conducted from 1957–1963. A network of 24 ground-based silver iodide generators comprised the nucleus delivery system. A statistical analysis of the days with hail indicated that "seeding appeared to be effective in increasing the number of hail days." Grossversuch IV (1977–1981) followed, in which Russian rockets were used in an attempt to replicate the striking successes reported from various republics of the U.S.S.R. The evaluation of Grossversuch IV did not confirm the Russian results. Some Russian scientists stated that the operational seeding on Grossversuch IV failed to follow the actual Russian procedures (Federer et al. 1986).

Approximately two dozen research entities in the United States were represented in Project Hailswath during the summer of 1966. Project Hailswath was a field research program headquartered at Rapid City, South Dakota. There were interesting indications of seeding effects, but the scientists involved agreed that firm conclusions were not possible from such a short-term project.

The establishment of the National Hail Research Experiment (NHRE) in northeast Colorado, under the guidance of the National Center for Atmospheric Research (NCAR), was a step toward an improved design for a hail suppression research program. The NHRE included both a randomized seeding experiment, which was in the field from 1972 to 1974, and a broad program of physical research (Foote and Knight 1977, 1979). Small, spin-stabilized rockets fired upward from airplanes were the primary source of silver iodide nuclei. The primary statistical evaluation declared that "no change (statistically) significant at the 10% level in the distribution ascribable to seeding could be detected." However, a later analysis of the distribution of seeding material in precipitation at the ground showed that much of the precipitation from seeded storms contained no silver iodide (Linkletter and Warburton 1977, Warburton et al. 1982). This finding suggested that the seeding agents had not been released in the proper locations in and around many of the "seeded" storms. The evaluation of the NHRE efforts was also made more difficult by changes in the target area. At the time, the instrumentation required to measure much of the physical evidence (e.g., polarimetric

radar) did not exist. A good general review of the status of hail suppression as of the late 1970s can be found in Dennis (1977).

The use of hail cannons has seen a resurgence in recent years. The physical principles upon which such instruments rely have not been published in the refereed scientific literature, and assessments of the technique suggest no suppression effect (Mezeix et al. 1974, Changnon and Ivens 1981). Perhaps most interesting is that every theory espoused regarding how a hail cannon might work is based upon the prevention of hailstone formation; yet the cannons always point directly upward, while the hailstorm approaches from a distance of many miles, generating large hail aloft as it moves toward the hail cannon. Hailstorms are very seldom stationary. Therefore, the trajectory of a hail embryo as it grows into a large hailstone and begins its final descent to the ground covers many miles in the horizontal before it reaches the ground. Only during the last few minutes of the (now large) hailstones' final descent to the ground is there any possibility of them being above or nearly above the hail cannon—and no means is known by which such large stones might be broken up. The shock wave from the explosion in the hail cannon propagates perhaps only 25 m (82 ft) from its origin (List 1963), and though the acoustical (sound) waves travel much farther, there is no frequency produced by the cannon that is not naturally produced by thunder. Therefore, this document does not consider hail cannon to be a viable means for hail suppression, and does not further address this technique.

1.2 THE STATUS OF HAIL SUPPRESSION TECHNOLOGY

A general discussion of the status of hail suppression technology as of the mid-1990s is provided in a summary document produced by the World Meteorological Organization (WMO 1996). Because our knowledge about hail formation is incomplete, there is still an experimental nature to present hail suppression operations. This should not be discouraging. There is, to some degree, a continuing experimental aspect to any complex activity conducted on a daily basis (compare, for example, with medical practices). In the case of hail suppression, the fact that we have difficulty in fully explaining many of the atmospheric phenomena that interact to produce hail on the ground accounts in large part for the lack of precision in estimates of the effectiveness of hail suppression technology. Nevertheless, operational hail suppression programs continue. At present, there are approximately 17 countries where serious hail suppression operational projects are in progress.

The participants of a recent WMO-sponsored meeting noted several new developments in remote sensing, computing capability, *in situ* measurements, and production and delivery of seeding agents, which suggested that further advances in the technology are now possible (WMO 1996). The experts concluded that the tools now exist to provide answers to many of the long-standing questions about hail formation, and urged that research programs and ongoing operational programs use as many of these tools as possible to quantify the effects of seeding and to elucidate hail formation processes.

The following policy and/or capability statements summarize present opinions of the scientific establishment on the subject of hail suppression.

1.2.1 American Society of Civil Engineers

The American Society of Civil Engineers, in Manual No. 81 (Kahan et al. 1995), states the following:

> *Several hypotheses have been advanced for the suppression of hail using artificial ice nuclei. A few of them may be termed, respectively, the beneficial competition hypothesis, the embryo competition hypothesis, and the premature rainout hypothesis. More elaborate hypotheses have been formulated on some projects. These hypotheses relate to the dynamics of the storm as well as to the microphysical growth processes of the hail embryos and hailstones.*

1.2.2 Weather Modification Association

The most recent hail suppression capability statement adopted by the Weather Modification Association (WMA 1986) reads as follows:

> *Most of what is currently known about the status of hail suppression, either success or failure, has been acquired from the study of surface hail data in a project area during seeding periods. Little has yet been shown through careful study of the physical behavior of the interior of storms from the suppression efforts. Therefore, the scientific linkages establishing hail suppression are not well established, although the assessment of surface hail differences are generally suggestive of successful suppression in the realm of 20–50% reduction. Execution of the operations is important. Timing and correct placement of seeding material are especially critical to successful suppression.*

1.2.3 American Meteorological Society

The most recent hail suppression capability statement adopted by the American Meteorological Society (AMS 1998) reads as follows:

> *Results of various operational and experimental projects and numerical modeling experiments provide a range of outcomes: some suggest decreases or increases in hail while others have produced inconclusive results. Statistical assessments of certain operational projects indicate successful reduction of crop-hail damage, but the physical basis for these results has not been established.*

1.2.4 World Meteorological Organization

The current WMO capability statement on hail suppression (WMO 1992) reads as follows:

> *Hail causes considerable damage to crops and property. Many hypotheses have been proposed for suppressing hail. The most common rationale has been creating enhanced competition among hailstones and their embryos for available moisture. This hypothesis is known as "beneficial competition." More recently efforts have been focused upon (in addition to the above) glaciating as much of the supercooled water as possible in the cells fueling the major hail-producing updrafts.*
>
> *Such seeding strategies may lead to premature rainout of the "feeder" cell and/or produce more ice in the cell, essentially robbing the hail-producing updraft of additional energy.*
>
> *Seeding clouds for such effects concentrates on the peripheral regions of large storm systems. It is generally ineffective to seed directly into the main hail-forming updraft as in most cases the nuclei would end up as small ice crystals in the anvil.*
>
> *At times the feeder cells themselves may have the potential to develop into a mature storm capable of producing hail. The above seeding method may then produce early precipitation and less hail.*
>
> *Only recently have numerical cloud model simulations been carried out to test these concepts. They do in general support these general concepts and also illustrate the*

FIGURE 2-1. Hailstorm, top view, Alberta, Canada. (Photograph by Terry Krauss, Weather Modification, Inc.)

important interactions that can occur between the seeded cells and the mature storm. Unfavorable interactions can lead to increase in hail with decrease in rain. The actual situation is normally very complex but competent operational and research scientists are working hard to delimit the favorable times, locations, and seeding amounts for effective modification treatments. Much more research is needed in all phases of hail suppression.

2.0 HAIL CONCEPTS

Hail, being the product of vigorous deep convection, requires those same conditions that are conducive to the development of thunderstorms (see Figures 2-1, 2-2). These include water vapor, especially in the low levels, instability (warm air in the low levels, with cold air aloft), and a trigger mechanism: meteorological conditions that will initiate storm development.

Recent advances in technology have provided and continue to provide an increasing convective storm knowledge-base. The physical concepts upon which hail suppression operations are based are subject to refinement and modification as well, but are summarized as follows.

2.1 REQUIREMENTS FOR HAIL DEVELOPMENT

A hailstone is formed whenever a frozen particle in a cloud encounters supercooled water in an updraft structure that keeps the nascent hailstone aloft within the cloud during its growth stage. A frozen particle that has the potential to grow into a hailstone is called a

FIGURE 2-2. A ground view of a hailstorm, North Dakota. (Photograph by Bruce Boe, North Dakota Atmospheric Resource Board.)

"hail embryo." An embryo can be an ice crystal, a frozen cloud droplet, a piece of a larger ice particle, or a frozen raindrop. Many hailstones melt into rain during their fall to earth; therefore, hail at the ground is more prevalent in mountains or high plains than it is at places near sea level. However, hailstones greater than about 2 cm (0.8 in.) diameter will reach the ground without completely melting under almost any conditions in cloud and below cloud base.

The combination of updrafts and supercooled water concentrations required for hailstone formation is found only in vigorous convective clouds. The severity of hailstorms is affected by many interacting factors, including the static stability of the atmosphere (influenced by the time of day), terrain, supply of water vapor near the ground, and the winds at all levels of the troposphere. The proximity of strong jet stream winds appears to favor intense convective developments, and hence the probability of large hail.

Hail mostly ravages those regions of the world that are located in middle latitudes downwind of mountain ranges and have ground elevations of 500 to 3,000 m (1,600 to 9,800 ft) above sea level. They include the regions in the lee of the Rocky Mountains from Alberta to northern Mexico, in the lee of the Andes in western Argentina, near the Pyrenees, the Alps, and the Caucasus Mountains in Europe, and near the Tibetan plateau in western China.

The smallest hailstorms may consist of a single convective cell with a lifetime of 20 or 30 min. However, most hailstorms are multicellular and may last for several hours, while individual cells form and dissipate within them. Hailstorms arranged in squall line form may last even longer, especially if the squall line is associated with an active cold front. Some storms evolve into "supercells," which may last several hours in relative isolation. These supercell storms are characterized by quasi–steady-state updrafts which appear to ingest adjacent developing cloud turrets, as compared to multicell storms in which the individual flanking line turrets each in turn grow, mature, and then rain out as they dissipate. Supercell storms account for a large fraction of all very large (5 cm [2 in.] diameter and larger) hail and the majority of damaging tornadoes. Because they are among the most intense storms, their successful modification is also more problematic.

Severe hailstorms usually extend to the tropopause, and sometimes penetrate a kilometer or more into the stratosphere. As the height of the tropopause decreases toward the poles, the height of a typical hailstorm varies with latitude. Hailstorms in Texas often tower to 15 to 18 km (49,000 to 59,000 ft) above sea level, while a hailstorm in Alberta might

reach only to 8 or 10 km (26,000 to 33,000 ft). In either case, updrafts of 10 to 15 m s^{-1} (22 to 34 mi h^{-1}) are common, updrafts in stronger storms easily exceed 30 m s^{-1} (70 mi h^{-1}), and the strongest updrafts that produce giant hailstones approach 50 m s^{-1} (110 mi h^{-1}) (Musil et al. 1986). While most hailstones are relatively small (~ 1 cm [0.4 in.] diameter), larger hail is common and especially damaging to property. An extremely large hailstone is shown in Figure 2-3.

The formation and growth mechanisms of hailstones are complex. The supercooled water that is the raw material for hailstone growth may exist as supercooled cloud droplets or supercooled rain and drizzle drops. The release of latent heat as the supercooled water freezes onto the growing hailstone warms the hailstone surface, often to 0°C. Hailstone growth in this situation is called "wet growth." Wet growth is limited by how quickly the hailstone can shed this heat to the much colder ambient air, and by how rapidly excess water is shed from the surface of the wet hailstone, slowing its growth. When there is less supercooled liquid water or when the growth environment is very cold, the latent heat release is insufficient to warm the hailstone to 0°C, and supercooled water freezes very quickly on contact, a condition referred to as "dry growth." In clouds with high concentrations of supercooled water, the transition from wet to dry growth occurs at surprisingly low ambient temperatures, sometimes down to −30°C. It is in strong updrafts at temperatures of −20 to −40°C that hailstones can grow most rapidly, a fact that explains the association of hail with tall, vigorous convective clouds.

2.2 THE SCIENTIFIC BASIS FOR HAIL SUPPRESSION

The precise conditions that make one thunderstorm produce hail while others do not remain a matter of some conjecture. Every vigorous thunderstorm contains millions of particles that could function as hail embryos, yet many thunderstorms produce negligible hail, or no hail at the ground.

Embryos that grow into large hailstones must encounter ever-stronger updrafts that support them inside the cloud as they grow. This steadily strengthening updraft may be due to the intensification of the host convective cell, or it may be realized for a particular embryo by a fortuitous trajectory that leads it into the more intense parts of a storm as it grows. Hail suppression, to be successful, must disrupt this complex process, because there is no known way to break up a hailstone in mid-air once it has formed.

FIGURE 2-3. Hailstone from Coffeyville, Kansas. Hen's egg, shown for scale in lower left, is about 6 cm (2.3 in.) in length. (Photograph courtesy of Nancy C. Knight, National Center for Atmospheric Research.)

The effects of seeding clouds for hail suppression on the storm's rainfall are not fully understood. Some evaluations from nonrandomized, operational hail suppression programs suggest net increases in rainfall, while others indicate no discernible impact. Statistically significant changes in rainfall have not been reported in the literature. Nevertheless, it is a natural concern for those near and within project target areas, so means to collect and analyze meaningful precipitation data should be considered. Numerical cloud modeling can be useful in this regard (Farley et al. 1996).

A number of hail suppression concepts have been formulated. The five concepts most commonly employed in the context of operational hail suppression projects are summarized below.

2.2.1 Beneficial Competition

Adding to the supply of hail embryos naturally available in theory could lead to the formation of more numerous but smaller hailstones. For complete suppression of hail, the concentration of hail embryos must be increased to the point that the more numerous embryos "compete" for the available supercooled liquid water, resulting in hailstones that are sufficiently small that they melt entirely during their descent through the warm subcloud layer. The risk associated with the beneficial competition concept is that one might add too few additional ice particles and simply produce more hail. Alternatively, the resulting hailstones might be more numerous and somewhat smaller, but still large enough to survive transit through the warm subcloud melting layer. In either case, hail will still reach the ground. In many cases, increased numbers of smaller hailstones might effectively reduce property damage, but a diminution of crop-hail damage is less certain.

2.2.2 Early Rainout

This concept calls for treatment(s) that speed the development of precipitation within treated cumulus congestus cells. Ice-phase precipitation thus develops faster as a result of treatment, falling from the clouds and melting before experiencing the stronger updrafts that would sustain hail growth. This early rainout of precipitation results in the precipitation of smaller hydrometeors from less mature, less vigorous clouds.

2.2.3 Trajectory Lowering

If the developing hydrometeors do not fall from the cloud complex (as in early rainout), they may still grow sufficiently large to follow a lower trajectory (shorter growth and residence times) so that they do not get carried aloft into the greatest supercooled water region of the storm during their lifetime. As a result, their size will be limited and they may still be able to melt completely before reaching the ground. Trajectory lowering might be the result of attempts to achieve early rainout; the two concepts are closely related.

2.2.4 Promotion of Coalescence

The promotion of coalescence of cloud droplets near cloud base can affect the microphysics of an incipient hailstorm in a number of ways. It may lead to early rainout, with associated trajectory lowering, and reduce the flux of water to the upper supercooled cloud. It also may provide frozen drops as an enhanced hail embryo source for beneficial competition. The promotion of coalescence is accomplished not by glaciogenic seeding but through treatment with hygroscopic materials.

2.2.5 Dynamic Effects

The release of convective instability in numerous small cloud towers over a larger area, rather than within a single intense, hail-producing updraft, could conceivably reduce the damaging hail from a large storm. Some have hypothesized that downdrafts induced by seeding might disrupt the organization of the strongest updraft, or generate outflow boundaries that trigger new convection.

2.2.6 Other Concepts

A sixth concept, glaciation of supercooled cloud water in the main updraft, has fallen from favor, largely because of the very large amounts of seeding agent required to achieve complete glaciation. If the seeding agent used relies primarily upon contact nucleation (the chance encounter of ice nuclei [IN] with supercooled liquid water drops), timing is also a factor. In such cases, when nucleation of cloud droplets is to occur primarily through collisions between the droplets and nuclei, 10 or 20 minutes might be required unless tremendous numbers of nuclei are introduced. In that time period, nuclei introduced into a strong updraft at cloud base would be ejected from cloud top without ever encountering a cloud droplet. A number of modern seeding agent formulations produce ice nuclei that have hygroscopic (water-attracting) tendencies, and these formulations function in the much faster condensation-freezing nucleation mode. However, even if nucleation is achieved in the mature updraft as soon as the seeding agent reaches supercooled altitudes, the time required for the resulting ice crystals to grow by diffusional growth (about $1 \ \mu m \ s^{-1}$) to precipitable sizes would result most often in increased ice concentrations, primarily within the cloud anvil. The greatest difficulty may reside in glaciation of an updraft of many kilometers diameter. Adequate dispersion of sufficient nucleating agent within such a very large volume would be a daunting task.

2.3 CONCEPT VISUALIZATION

Figure 2-4 depicts a multicell thunderstorm, consisting of developing cloud turrets organized in a "flanking line" (left), a mature cell possessing a strong updraft and large (and still growing) hydrometeors, and dissipating cells producing precipitation (right). In the archetypal supercell, the principal defining characteristic is a quasi–steady-state mature updraft, which ingests the growing clouds of the flanking line as they mature. In multicellular storms, the developing cells of the flanking line each in turn mature, becoming the dominant cell, as the older cells weaken and rain out. No differentiation between the two storm types is made in Figure 2-4; therefore, the concepts presented below may apply to both. Figure 2-4 illustrates hail

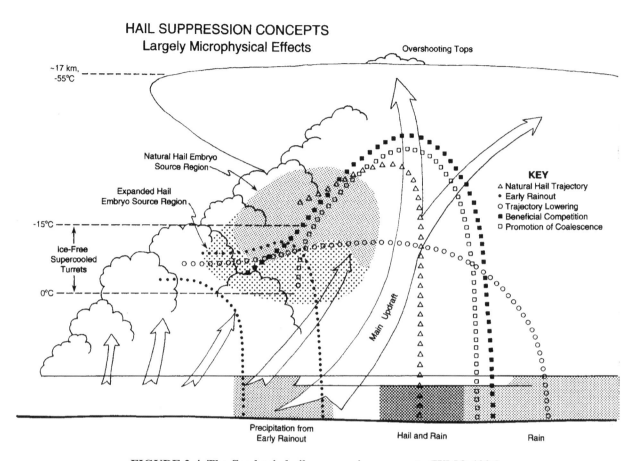

FIGURE 2-4. The five basic hail suppression concepts (WMO 1996).

suppression concepts based primarily on microphysical effects from seeding, as described in Sections 2.2.1 through 2.2.4.

The five concepts share the following:

1. Developing cloud turrets are treated, probably cumulus congestus, rather than cumulonimbus. This means treatment of young clouds with modest updrafts, rather than mature cells and strong updrafts.
2. Natural precipitation development is accelerated. Promotion of coalescence is initially directed at liquid-phase processes; the others are based largely on glaciogenic effects. Dynamic effects result from the release of latent heat (primarily from freezing), and from redistribution of condensed water within the targeted cloud turrets.

The dominant path of development of hail in unseeded (natural) storms is shown in Figure 2-4 by the open triangles ($\triangle$). The path originates in the central portion of the natural hail embryo source region, at temperatures somewhat colder than $-15°C$. The embryos are carried far aloft as they become hailstones

within the strong, mature updraft. As they reach the edge of the updraft or gain terminal velocities exceeding those of the updrafts, descent begins, and a hail cascade develops near and beneath the sloped updraft.

Timely seeding of growing, supercooled, largely ice-free turrets on the storm flank is illustrated in four ways, all somewhat interrelated.

The early rainout concept is illustrated by the paths shown by the solid dots ($\bullet$). Two of the many possible trajectories are shown, but the objective is clear: get a significant fraction of the developing hydrometeors to precipitate before encountering the main updraft, where hail development might occur. The "early" precipitation ideally would carry a fraction of the supercooled cloud water with it as well, reducing the supply remaining to support hail growth.

The trajectory lowering concept ($\bigcirc$) is illustrated as an extension of the early rainout concept. In this case, early rainout is not achieved, but the hydrometeor development has been sufficiently advanced that the natural hail embryo source region is largely avoided, and the resulting graupel and/or small hail melts before reaching the surface. The hydrometeors must be large enough to fall from the cloud before a mature updraft lifts them into a supercooled hail growth environment.

The accelerated hydrometeor development resulting from seeding may produce large numbers of potential hail embryos, many or most at altitudes below the natural hail embryo source region. This expanded hail embryo source region provides the initiation of beneficial competition, as shown by the solid squares (■). With this concept, the idea is to redistribute the ice mass spectrum from low concentrations of large hailstones to high concentrations of small hailstones. The small hailstones have a much greater probability of melting before reaching the ground. When this spectral change is achieved, the smaller stones may be carried even higher, and thus the trajectory may be raised. This may not always be the case, however, for if the total ice mass was significantly increased, the additional mass loading could act to slow the updraft.

The promotion of coalescence is an alternate means of effectively inducing beneficial competition; hence the trajectory (□) is shown approximating that of the latter. Hygroscopic (rather than glaciogenic) seeding acts to modify the droplet spectrum, resulting in greater numbers of large droplets, which thus activate the coalescence process. Since the ice-phase processes are not initially involved, and because greater cloud water contents (as produced by stronger updrafts) aid this process, the source region may be closer to the stronger updrafts, and perhaps lower in the cloud. Ideally, the supercooled raindrops freeze, creating a ready and ample supply of embryos for beneficial competition. Ice multiplication through the splintering of freezing raindrops also may contribute to glaciation.

Dynamic effects are not specifically illustrated in Figure 2-4. The most common of these would be additional growth of the seeded turrets (latent heat release due to early glaciation), increased updraft speeds within them, and the development of incident downdrafts and outflows associated with the accelerated development of precipitation. One might also plausibly argue that release of greater convective instability within the flanking line would effectively redistribute it over a larger area, thus perhaps actually weakening future updrafts as they mature.

2.4 CLOUD MODELING

Considerable success has been achieved using numerical cloud models to simulate hailstorms and hail development (Farley et al. 1996, Orville 1996). Cloud modeling has contributed significantly to the develop-

ment of the contemporary conceptual models for hail suppression. Contemporary models now contain microphysical parameterizations, as well as cloud dynamics.

If a cloud model, after "programming" with actual atmospheric conditions, can successfully reproduce a cloud or storm like that actually observed in those same atmospheric conditions, the model has reinforced the physical concepts employed therein. Put another way, cloud models are tools that can be used to test concepts. If the model employing a certain concept gets it right, this strengthens the faith in the concept. (For more information on numerical cloud models and their application, see Section 3.4.5.)

3.0 THE DESIGN OF HAIL SUPPRESSION OPERATIONS

When hail suppression operations are being considered, it is already understood that hail damage is a significant problem in an area in which the political subdivision, organization, or other sponsoring entity has control and/or responsibility. The next step is to determine the scope of the hail problem, the geographical area requiring protection, and the seeding agents and methods to be employed. Consideration also must be given to how the project is to be evaluated. Once these tasks are done, the framework of the program can be established and a project design produced. These basics are discussed in this section.

Planners are urged to consult legal counsel to carefully weigh all aspects of their proposed program in the context of applicable local law. The sponsors also are encouraged to engage a scientific team knowledgeable of severe storms and cloud seeding, to either develop or review the design and intent of the proposed operational program, and to render opinions as to the likely efficacy and safety of the proposed program.

3.1 DEFINITION OF PROJECT SCOPE

The project target area generally is the geographical area or portion thereof in which the sponsoring entity has an interest. If the project is to reduce property damage, the focus will be on areas in which property is concentrated, i.e., cities and towns. If reduction of crop-hail damage is the primary function, the target usually will be significantly larger.

The definition of the target area and the method(s) for delivering the seeding material are among the most important decisions to be made in the design phase. When designing a new project, all information available about the local area and the proposed targeting and delivery methods should be analyzed in depth. This includes the climatology, meteorology, and geography of the target area; the seeding techniques and equipment to be employed; and likely operational procedures. In addition, the incorporation of data collection efforts that will allow project evaluation should be integrated into the overall design. Evaluation considerations are discussed in Section 5.0.

It is essential that local legal requirements also be considered. Most areas require at minimum the licensing and permitting of operations. Others may have minimum qualifications for those individuals actually conducting the operations. Environmental regulations also may require some additional study prior to operations. Legal and environmental requirements are discussed further in Sections 3.6 and 3.7.

3.1.1 Historical Hail Losses

The aim of hail suppression projects is to reduce the losses caused by hail. Finding a reliable source of historical data concerning the losses in the project target and control areas over a long period will be helpful in establishing economic objectives and for evaluating the resulting economic impact of the hail suppression efforts.

Hail damage affects a range of economic activities, although hail suppression projects are typically operated in areas where agricultural interests dominate the local economy. Unfortunately, hailfall in most areas tends to occur mainly during the growing seasons. Because many factors are involved in the assessment of crop damage, obtaining a reliable database on the percentage of losses caused by hailfall is not an easy task. For example, the magnitude of the crop damage is related to hailfall intensity, crop type, and the growth stage and health of the plant. Even so, some kind of database of past losses should be developed, if possible, given that it will allow economic objectives to be more realistically established.

Some hail suppression projects are now focusing on urban areas to reduce property losses (Krauss and Renick 1997). These projects are not limited only to the growing seasons but extend to all times when hailstorms might develop. In Alberta, this translates into an earlier start date. Potentially damaging hail occurs there in early summer before crops have begun to develop, and property (homes, automobiles, etc.) is exposed and subject to damages. The construction of evaluation models based on simulated results can assist the establishment of such objectives (Sánchez et al. 1996, Hohl et al. 2002).

3.1.2 Historical Radar Data

In many locations, historical radar data may be obtainable. These data often will reveal preferred areas for storm genesis, common storm motions and paths, and, very often, detail about the cellular character of the storms themselves. Treatment strategy for storms may vary depending upon whether the storms are multicellular in character or, more often, are characterized by a single, persistent, long-lived mature updraft (supercells). In the United States, the best source for such radar data is the National Climatic Data Center (Asheville, North Carolina), which archives the National Weather Service WSR-88D (NEXRAD) Doppler weather radar data for all sites.

3.1.3 Basic Project Area Concepts

Concepts relating to the target area, operations, and evaluations are stated below. These terms, which are fundamental to the design of any hail suppression project, are illustrated in Figure 3-1.

1. **Target Area.** The area in which the objective is to reduce the losses/damages due to hail. Seeding is conducted with the intent of affecting hailfalls over the target area.
2. **Control Area.** An area over which unseeded cloud systems are observed for comparisons with those clouds over the target area. It must be located near the target area and be of similar size and climatology.
3. **Operational Area.** The area over or adjacent to the target area within which seeding operations are actually conducted. The operational area ideally should extend upwind of the target area because the effects of seeding are not instantaneous. Clouds that develop upwind of the target area must be treated before they move over it, as well as while they are above it.
4. **Buffer Zone.** The area separating the target area from the control area; buffers the control area from the possible effects of seeding within the operational area.

3.1.4 Initial Design Considerations

After the project area has been defined, the other basic parameters can be defined. This is done by answering the following questions:

1. During what seasons/months is the project scheduled to operate?
2. What seeding method(s) is (are) best suited for the project? Will seeding be from the ground, from aircraft, by rocket, or by some combination? If by aircraft, will it be subcloud in updraft, or by direct injection of the seeding agent(s) at cloud top, or both?

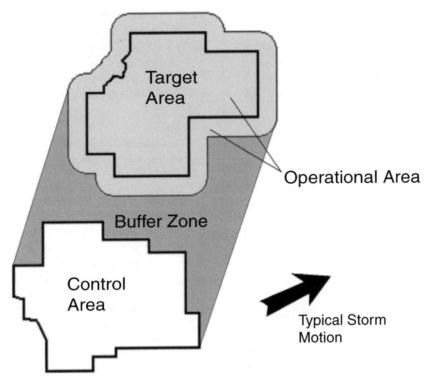

FIGURE 3-1. Terms related to the project area and operations.

3. Will the character of the seeding agents be glacio-genic, hygroscopic, or both, and by what mechanism will the agents be expected to function?
4. Will operations be conducted around the clock, or only during daylight hours?
5. How will operations be directed?
6. What will be the role of forecasting, both short-term and long-term?
7. What types of meteorological data will be available, and how will they be utilized?
8. How will the project effectiveness be evaluated?

Before final decisions are made, it is recommended that all seeding options be appraised and that both economic and scientific perspectives be carefully weighed. Once the design has been determined and the project established, the basic design should not be altered unless unresolvable difficulties are immediately discovered. Doing so will complicate evaluation efforts, which, because of natural variability in the weather, cannot be based on just one or two seasons. This reemphasizes the importance of a sound initial design.

3.1.5 Climatology

The most complete climatological information available should be collected about the target and control areas. The WMO (1992) recommends the acquisition of climatological data during the design phase. This knowledge of storm and hail climatology should be incorporated in the design of a project. Many references provide useful information about hailfalls and hailstorms through climatological analysis of different parts of the world. See, for example, Boe 1992, Carte et al. 1963, Changnon 1977, Dessens 1986, 1998, Dye et al. 1976, Federer 1977, Foote and Knight 1977, 1979, Hitschfeld 1971, Morgan 1973, 1982a, 1982b, Morgan et al. 1980, Prodi 1974, Sánchez et al. 1994, and Smith et al. 1997.

Project designers must consider the following climatological variables:

1. Surface area affected by typical hailfalls
2. Range of hailstone sizes observed
3. Diurnal distribution of hail
4. Distribution and locations of known high risk areas; those which tend to suffer from hailfall most frequently
5. Monthly or seasonal distribution of hailfalls
6. Spatial and temporal distribution of the appearance of radar echoes from convective clouds
7. Concentration of natural ice nuclei
8. Relationships between radar-derived storm characteristics and hailfall at ground level

Design of the project will be further enhanced if, in addition to these variables, information is available about the general behavior of convective cells in the project area, the expected range of hailfall intensities, and the relative quantities of water and hail that are typically precipitated.

Possible local sources of real-time meteorological data should be located to support project forecasting and nowcasting. The Internet is one possibility; local colleges, universities, and governmental weather offices are others.

Climatological analyses should identify areas prone to storm development, which should govern the placement of seeding facilities and also aid operational decision-making.

3.2 DELIVERY METHODS

The consistent, proper placement of the seeding agent(s) in both time and location (targeting) is absolutely essential in any program of applied cloud seeding and cannot be overemphasized. Many programs have failed to produce the desired results, only to later find in the course of post-analyses that the seeding agent was not being properly targeted.

It is therefore extremely important that a delivery method be selected that will function reliably in the meteorological conditions and storm environment typical of the project. In this section, an overview of the various equipment required for the generation of the seeding aerosols and delivery of the seeding agents is provided, along with illustrations of each type.

There are two basic philosophies currently applied in operational hail suppression programs: *broadcast seeding* and *direct targeting*. Broadcast seeding is the continuous emission of seeding agent (usually silver iodide (AgI) aerosols), most often from ground-based generators. This significantly increases the concentration of ice nuclei in the atmosphere when forecasts suggest that there is a risk of hail in some part of the target area. AgI aerosol generators are positioned with the intention of boosting the ice nuclei concentration in and below the hail growth zone, so that, when the hailstorms are forming, the inflow to the storms includes a portion of the boundary layer volume containing the aerosol.

Direct targeting involves the application of seeding material directly to the cloud, at the place and moment when the mechanisms that govern the formation and growth of hail may be successfully modified or restrained. Means of detecting developing storms early

and of predicting the risk of hail are, therefore, required for both broadcast seeding and direct targeting to be effective.

Around the world, there are many hail suppression projects currently operating or under development. The WMO periodically edits and publishes the *Register of National Weather Modification Projects* (WMO 1997), which gives an account of most of these projects and of the systems they use. In these projects, the delivery of seeding agent is carried out by one or more of the following means:

1. Aircraft carrying seeding devices that may be nuclei generators that burn an AgI-complex-acetone solution, pyrotechnic flares, or mechanisms that dispense dry ice. This delivery technique is currently in widespread application, and is favored for its relatively certain targeting and modest response time.
2. Ground-based ice nuclei generators, usually in networks that emit continuous plumes of AgI aerosols. This is the type of system used in southern France since 1951, across wide areas of Spain since 1975, and in various other countries around the Mediterranean Sea. This delivery technique requires a longer response time and often uses more seeding agent. It relies upon local convective air currents to transport the seeding agents to the growing clouds.
3. Rockets or artillery shells containing nucleating material. These systems have been widely used in the past (see Section 3.2.3). In spite of the very small response time required with rocket systems, in recent years rockets have been employed less often, probably due to the expense of the rockets themselves and to restrictions upon their use imposed in many locales.

As there may be no "best" delivery technique, the possibility of using a mixed system should not be ruled out. One system might be more suitable than another, for instance, during day or night situations or over a particular type of terrain.

3.2.1 Airborne Application

Ice nucleus generators located under the wings or on the wing tips, racks of ejectable flares mounted on the bottom of the aircraft fuselage, burn-in-place flares held in wing racks, and dry ice dispensed from a hopper internal to the aircraft all can be used to seed effectively in airborne applications (Figures 3-2, 3-3). In the case of the generators, the aerosol emission is of the order of 100 to 200 g (0.22 to 0.44 lb) AgI per hour, while the burn-in-place flares, which burn from one to several minutes, typically contain from 20 to 150 g

FIGURE 3-2. Cloud base seeding aircraft with wing-tip ice nuclei generators (inset). The generator shown is pressurized by ram air and has a 30 L (8 gal) capacity, which allows continuous seeding for about 2.5 h at normal air speed. (Photograph by Bruce Boe, North Dakota Atmospheric Resource Board.)

FIGURE 3-3. A typical cloud-top seeding aircraft, equipped for seeding with dry ice, 20 g (0.04 lb) ejectable flares (inset, right), and wing-tip ice nuclei aerosol generators. Cloud-top aircraft must be able to operate at altitudes where modestly supercooled (−5 to −10°C) cloud tops are found; this is usually between 4.6 and 6.7 km (15,000 to 22,000 ft) msl. Though cabin pressurization is desirable, it is not an absolute requirement. The lower photograph shows a wing-tip generator type pressurized by ram air (see also Figure 3-2); another type having internal pressurization is shown in the upper left. The dry ice hopper is not apparent from the exterior of the aircraft. (Upper photographs courtesy of Terry Krauss, Weather Modification, Inc., lower image courtesy of Bruce Boe, North Dakota Atmospheric Resource Board.)

(0.044 to 0.33 lb). Some of the characteristics of these systems are described in ASCE Manual No. 81 (Kahan et al. 1995).

When dispensed at convective cloud base (from wing-tip generators or burn-in-place flares), the seeding material must be introduced into the ascending currents (updrafts) of the newly developing convective clouds on the flanks of the mature storms. This also is true of clouds that are developing independent of mature storms but are likely to develop into thunderstorms themselves. The aircraft typically has to be able to sustain flight in the subcloud updraft region, feeding the nucleant directly into the cloud updraft. While the targeting of this technique is relatively certain, time (perhaps 10 to 20 min) is still required for the seeding aerosol to be transported aloft to the supercooled portions of the cloud. When cloud bases are found at lower altitudes (warmer temperatures), the transport to supercooled cloud regions requires more time, so it is imperative that seeding be initiated sufficiently early.

This means initiation of treatment before the subject cloud tower becomes supercooled in most cases.

Seeding in mature updrafts should be avoided, as these updrafts often exceed 20 m s^{-1} (45 mi h^{-1}), and any ice that might result has no chance of growing to precipitable size before being transported to cloud top (see Section 2.2.5). In addition, such strong updrafts can and frequently do produce large hail, which is a risk to both aircraft and occupants. During direct injection of seeding agents from aircraft into the tops of growing supercooled turrets with ejectable flares or dry ice, vertical wind speeds (both rising and descending) of 10 m s^{-1} (22 mi h^{-1}) or more may be encountered. Strong turbulence is not unusual, especially in the shear zone between the descending air at the turret boundaries and the interior updrafts. When ejectable flares or dry ice are used, the seeding agent(s) can be released during penetration, or dropped into the cloud top from above, and a significant quantity of seeding material can be discharged in only a few seconds. The effects of this kind of delivery are almost immediate if the targeted cloud is supercooled.

The dry ice pellets must be small enough to sublimate as much as possible during their fall through supercooled cloud. American projects typically employ pellets of 11 mm diameter (0.43 in.). If released at cloud top in turrets growing upward at 10 m s^{-1} (2,000 ft per min), through the $-10°C$ ($+14°F$) level, a portion of each pellet remains still unsublimated by the time it falls into warm ($>0°C$, 32°F) cloud, and thus is wasted. Smaller diameter pellets may be used if available, as long as they are not so small as to clog the dispensing mechanism. Initial dry ice seeding rates during flight through supercooled updraft should be near 0.5 kg ($\sim$1.1 lbs) km^{-1} (Holroyd et al. 1978), but trials in the local conditions are recommended.

Ejectable flares that burn 20 g (0.044 lb) of nucleant in about 30 s are most often used for on-top seeding. Formulations producing nuclei that act predominantly in the condensation-freezing mode are preferred since, in hail suppression operations, quick nucleation of the supercooled liquid water is necessary. Smaller turrets will have limited updraft and supercooled liquid water, and only a single flare will be needed. An additional flare is fired every several hundred meters in broader or faster-growing turrets, or every 4 to 5 s if flying at 67 to 100 m s^{-1} (150 to 225 mi h^{-1}). This seeding rate is based on turbulent mixing observations in actively growing cumulus, which showed overlapping seeding plumes within 250 s (4+ min) with this rate (Grandia et al. 1979), with enough ice nuclei to deplete the cloud liquid water within a 10 m s^{-1} (22 mi h^{-1}) updraft (Cooper and Marwitz 1980).

The purpose of this type of seeding is to start ice formation before it would naturally occur, to generate significantly increased ice concentrations as the targeted turret matures. Therefore, treatment must be conducted on developing cloud turrets that are both supercooled and ice-free. Once ice begins to develop, it quickly multiplies within the cloud turret, so treatment of turrets that already contain significant (natural) ice is unnecessary and a waste of precious time and resources. A bigger turret is usually closer to developing ice naturally, and the objective is to give each turret as much head start as possible.

Modern Global Positioning Systems (GPS) are a relatively inexpensive means of confidently determining the position of the aircraft and ground vehicles in the field. With a suitable radio link, the aircraft position can be transmitted to the ground radar site and automatically plotted on the radar screen. This offers another important source of documentation of aircraft cloud treatment operations. It is possible to automatically record exactly when and where treatment was carried out and show this in relation to the radar or other types of images. The recording of aircraft position with GPS is not linked to the radar antenna rotation, as it is with the older Identification, Friend or Foe (IFF) transponder-type positioning system. The information also contributes to operational safety. In operations with more than one aircraft airborne at any time, the positional information can be exchanged between aircraft and aid in maintaining separation or coordinating movements. From the aircraft flight tracks and the verbal reports via radio from the pilots, the targeting can be accorded considerable confidence, especially when the seeding agent is released directly in the cloud turret.

The airplanes used for direct injection seeding must be equipped with the instrumentation required for flight under instrument flight rules (IFR) conditions. In the United States, the Federal Aviation Administration (FAA) is required to certify all cloud-seeding aircraft, including the modifications made to accommodate equipment, and this certification brings with it certain restrictions and limitations. Generally, the private companies equipping hail suppression projects use twin-engine, turbo-charged aircraft. Speed and rate-of-climb must be considered, as the aircraft must be capable of quickly getting to rapidly evolving clouds. Twin-engine aircraft are desired for these performance reasons, and also for the ability to safely return to an airport should one of the engines malfunction.

3.2.2 Ground-Based Ice Nucleus Generator Networks

The vaporization of AgI aerosols may be carried out by a generator burning a solution of acetone, AgI, and other ingredients. This process is discussed in greater detail in Section 3.3. The most common method is to use generators in fixed surface positions, forming a network. The number and position of these should be calculated considering at a minimum the following factors:

1. Characteristics of the aerosol emission from type generators and seeding solution
2. Natural background concentration of naturally occurring ice nuclei in the target area
3. Most likely wind trajectories prevailing in conditions conducive to the formation and/or propagation of convective storms
4. Information about local atmospheric stability, including inversions that might trap the seeding agent within the boundary layer, thereby inhibiting dispersion of seeding material into areas where storm inflow is likely to tap it

One approach is to space the ground-based ice nuclei generators at a distance comparable to the diameter of the smaller convective cells, on the order of 8 km (5 mi), as was done in some Great Plains hail suppression programs in the 1950s. This would ensure that any developing storm would have a chance of ingesting aerosol from at least one ground-based generator.

Modeling of seeding agent dispersion (dispersion modeling) also should be conducted (Admirat 1972, Pasquill 1974, Bruintjes et al. 1995) to aid the determination of the number and placement of generators required to deliver sufficient ice nuclei to cumulus cloud altitudes. The generators are generally kept burning throughout the time the suitable clouds are within the target area. Over this time period, a considerable amount of seeding agent is consumed, usually far more than would have been used had the clouds been seeded directly. However, the expense of operating aircraft is saved.

When possible, ground-based generators should be situated in mountainous or hilly regions upwind of the target area, so that the wind and natural turbulence will loft the seeding material and assist dispersion. The network used in southern France is shown in Figure 3-4 (from Dessens 1998). To ensure targeting in as many meteorological situations as possible, the network is made up of 467 generators deployed over about 40,000 km^2 (15,400 mi^2). Thus, it is apparent that development of ground-based generator networks is not simply a matter of deploying a dozen or so generators.

If the primary storm formation areas are known (through meteorological radar data, for example), the IN aerosol concentration can be boosted by the deployment of additional generators in these areas (Castro et al. 1998). Ground-based generators can be switched on and off manually, each one by a designated responsible individual, or automatically from a

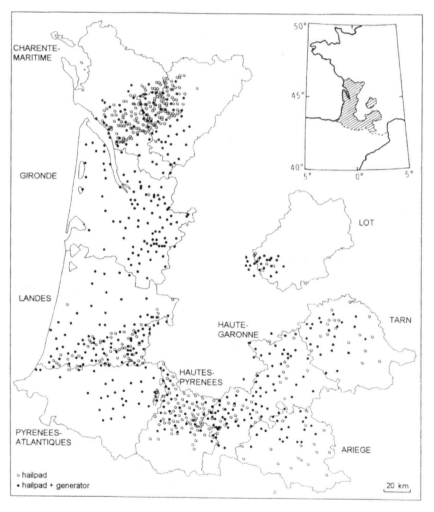

FIGURE 3-4. Map of the hail pad and seeding stations of the hail suppression program in southwestern France in 1995. (From Dessens 1998, copyright American Meteorological Society, used by permission.)

FIGURE 3-5. Ground-based ice nuclei generators. A and B, stand-alone units for remote use, which operate from two 24-v wet cell batteries charged by a solar unit. Communication is accomplished via satellite telephone. These units can use a wide range of silver iodide solutions and rates. Chlorinated acetone-based solutions of 1.5 to 2.0% AgI are typically burned at rates of 2 to 8 L (0.5 to 2.1 gal) per hour. C and D, manually operated generators typically placed near homes, where telephone calls to the residents provide the directives to turn on and off the generators. Solution concentrations and flow rates are similar to those for the remote generators. (Photographs courtesy of Thomas Henderson, Atmospherics, Inc.)

central control station. In both cases, maintenance service is essential to replace supplies when necessary and maintain each generator in operating condition. The aerosol emission level for ground generators typically varies from 15 to 40 g (0.033 to 0.088 lbs) AgI per hour. Figure 3-5 shows a generator, manufactured by Atmospherics Incorporated, that uses a propane flame to burn a 1.1% AgI acetone solution at the rate of about 2 L (0.53 gal) or 15 g (0.033 lbs) AgI per hour.

Given that at least some fraction of the generator network needs to release seeding material whenever there is a risk of storms, it is essential for the project to be able to count on accurate forecasting. This is particularly important when the generator network is not au-

tomatic. Because of the inevitable delay involved in switching on the generators; the individuals responsible for each machine have to be alerted and get to their posts before each generator can be activated. With an automatic network, however, the generators can be activated or deactivated within a few minutes.

Damaging hailstorms are not limited to hilly or mountainous terrain; the plains of North America regularly experience vigorous hailstorms. Therefore, it also may be desirable to deploy ground-based generators in relatively flat terrain, relying on convective heating to mix the seeding agent throughout the moist boundary layer upon which the developing storms feed (McPartland and Super 1978, Heimbach and Stone 1984, Griffith et al. 1990). While airborne targeting by aircraft flying in updrafts below convective cloud bases in visual meteorological conditions (VMC) has often been employed in North America, there are sometimes circumstances when excessively low cloud bases, darkness, or instrument meteorological conditions (IMC) make it unsafe to dispense seeding agent from below base. In such cases, an ancillary ground-based generator network might prove most helpful.

3.2.3 Rockets and Artillery

For many years, rockets and/or artillery have been employed to deliver seeding material into storm cells in China and the countries of the former Soviet Union (the Russian Federation, Moldavia, the Ukraine, and the Trans-Caucasian republics). Rockets also have been used in Argentina, Bulgaria, Hungary, Serbia, Croatia, Slovenia, and various other states. Neither rockets nor artillery have been employed in western countries because of the greater costs and the potential risks to aircraft and population.

Scientists of the former Soviet Union pioneered this hail suppression technique, which made use of rockets carrying substantial quantities of seeding material (several hundred grams of AgI each). Recently, the Russian Federation and others have developed the Antigrad Automatic Control System, which employs rockets to target developing supercooled turrets of the flanking line (Figure 3-6). When a convective storm meeting the radar criteria for seeding is detected, rockets are launched. One such rocket, the Alazan-5, can deliver 630 g (1.39 lbs) AgI and has a range of 9.5 km (5.9 mi). Another, the Krystal, can deliver 173.6 g (0.38 lb) AgI, in the form of 28 individual 6.2 g (0.014 lb) flares, to a maximum range of 12 km (7.5 mi). Both of these rockets are equipped with self-destruct charges to fragment the rocket after dispersion of the seeding agent. The newest rocket, the Alan, can deliver 42 g (0.093 lb) AgI, also to a maximum range of 12 km (7.5 mi). The Alan is parachute equipped; no self-destruct charge is needed.

The rockets are designed to release a plume of nucleating material once a certain height has been reached. Launch crews may receive details about the necessary elevation and azimuth angles and AgI ignition altitude for the launch from the radar, or the targeting may be automated by telemetry of aiming and firing instructions to the launcher. The greatest advantage of the system is its potential to reach any region of convective cells in a very short time; all of these rockets will attain their target within one minute.

Disadvantages are firing restrictions due to air traffic and, in some cases, the presence of populated areas. It is doubtful that such a system could be used in the United States, as the requisite federal approval almost certainly would not be obtained. In addition, the self-destruct mechanism occasionally may misfire, so care has to be taken to launch the rocket in a direction that will not result in damage to the public or to property. With the Antigrad system, such sectors can be programmed into the launchers, so that firing within them will not occur. A more serious shortcoming may be that the targeting instructions are derived from the radar returns, yet the desired ice-free, supercooled cloud turrets will rarely generate a discernible radar echo with radars typically used in operational hail suppression programs. Rockets and artillery are unacceptable as a means of seeding in urban areas.

The verification of the targeting for such rockets has never been accomplished. Though visual and radar indications suggest that the rockets can reach the intended cloud complexes, the release and dispersion of the seeding agent within the intended turrets has not been verified by in situ microphysical measurements.

Finally, there must be a sufficient number of launching sites to cover the seeding area. For an area of 1,000 km^2 (386 mi^2), from 20 to 25 sites are required, each one of which needs at least one launch pad and an arsenal for storing the rockets. It is notable that in each of these rocket-based projects personnel requirements are very significant, as there must be trained staff at the radar and launch sites whenever storms are forecast, sometimes 24 hours a day. All this makes it a very costly system to operate. When coupled with the expense of the seeding rockets, the potential for rocket debris to reach the ground after self-destruct or if self-destruct fails, and airspace limitations due to general and commercial aviation, most contemporary hail suppression projects have sought alternate treatment techniques. Anyone considering the use of rockets should first determine what local, state, and federal flight restrictions apply, and how such restrictions might impact operations.

3.3 SEEDING AGENT SELECTION

The materials placed within the targeted clouds are known as seeding agents. While glaciogenic agents intended to increase ice formation are the most common, others having hygroscopic properties are being used with increasing frequency. The full effects of this latter class of seeding agents are only beginning to be explored. Hail suppression involves intervening in the microphysical and/or dynamic development of convective cells in order to restrain, deactivate, or disrupt

FIGURE 3-6. The basic tools of hail suppression by rocket technology. A and B, the Russian MRL-5 radar is used to provide targeting information to the rocketry crews. Rockets presently available include Alazan (C, left), Krystal (C, right), and Alan (D). Automatic control system rocket launcher is depicted in (E); a launch is shown in (F). (Photographs by Aleksi Shachmurza, courtesy Magomet Abshaev, Antigrad Hail Suppression Research Center.)

those processes in the interior of convective cells that result in the formation of hailstones. Most of the time, rainfall is also desired, so whatever steps are taken to mitigate hail should not reduce rainfall but increase it.

The method most widely employed consists of introducing glaciogenic agents, materials that have the capacity to generate additional cloud ice. When added to the natural ice (if any) within the embryo source region, the collective cloud ice population should alter the cloud sufficiently to result in the suppression of damaging hail (see the discussion of hail suppression concepts presented in Section 2.2).

There are many natural glaciogenic agents. Not all, however, form ice crystals with the same facility, given that their efficiency in this respect is a function of their composition and size. In fact, each substance has a threshold temperature, which is the temperature at which it begins to cause the formation of ice crystals. The discovery by Vonnegut (1947) of AgI as an extremely efficient ice nucleating agent, with a threshold near $-5°C$ $(+23°F)$, was therefore a major contribution to weather modification activity.

Another method uses a quite different approach called "hygroscopic seeding" (Dennis and Koscielski

1972, Mather 1991). This technique aims to speed the development of large cloud droplets and raindrops through coalescence in the warmer (lower altitude) parts of the cloud. Such accelerated rain development may reduce the flux of liquid water into the higher, colder regions where hailstones form, or cause high concentrations of large droplets [>100 μm (0.004 in.) diameter] to enter the cold cloud regions. The principal mechanisms that hygroscopic seeding sets in motion are called "early rainout" or "beneficial competition" (see Section 2.2).

3.3.1 Silver Iodide

Silver iodide, in combination with various other chemicals, most often alkali iodide salts, has been used as a glaciogenic agent for half a century. In spite of its relatively high cost, it remains a favorite, especially in formulations which result in IN with hygroscopic tendencies.

Silver iodide has utility as an ice nucleant because it has the three properties required for field application: (1) it is a nucleant, regardless of mechanism; (2) it is relatively insoluble at $<10^{-9}$ g (2×10^{-12} lb) per gram of water, so that the particles can nucleate ice before they dissolve; and (3) it is stable enough at high temperatures to permit vaporization and recondensation to form large numbers of functional nuclei per gram of AgI burned (Finnegan 1998). Also, the ice crystallization temperature threshold for AgI is warmer than the threshold for most naturally occurring IN, which commonly have thresholds closer to $-15°C$ ($+5°F$). The chemical formulations of AgI seeding agents may be modified further, so that the resulting IN function at even warmer temperatures (DeMott 1991, Garvey 1975).

In many cases, AgI is released by a generator that vaporizes it (through combustion) and produces aerosols with particles of 0.1 to 0.01 μm (4×10^{-5} to 4×10^{-6} in.) diameter (see Figures 3-2, 3-3, 3-5, and 3-7). The process can be carried out using generators that burn a solution of acetone with an AgI concentration of the order of 1–2%. AgI is insoluble in acetone; commonly used solubilizing agents include ammonium iodide (NH_4I) and any of the alkali iodides. Additional oxidizers and additives commonly include ammonium perchlorate (NH_4ClO_4), sodium perchlorate ($NaClO_4$), and paradichlorobenzene ($C_6H_4Cl_2$). The relative amounts of such additives and oxidizers modulate the yield, nucleation mechanism, and ice crystal production rates.

The generation of AgI aerosols also can be accomplished by burning specialized pyrotechnics (Figures 3-8, 3-9). In many cases, a mixture containing silver iodate ($AgIO_3$) has been used. Powdered aluminum and magnesium, and some kind of organic agglutinant also are often added to the mixture (Dennis 1980). In recent years, advances in nucleation physics have resulted in a number of more effective pyrotechnic formulations that produce nuclei that, in addition to having ice nucleation thresholds near $-5°C$, also are somewhat hygroscopic. The resulting nuclei are not only effective as IN, but they also attract water molecules. This results in particles that in relative humidities near saturation quickly form droplets of their own, which then freeze shortly after becoming supercooled. This condensation-freezing nucleation process generally functions faster than simple AgI. Laboratory testing has shown that AgI by itself functions primarily by the contact nucleation process, which is more dependent upon cloud droplet concentration, and consequently a much slower process (DeMott 1991). Speed in nucleation is very desirable in applications such as hail suppression, where quick glaciation of modestly supercooled cloud turrets is required (DeMott et al. 1995, Finnegan et al. 1994, Pham Van Dihn 1973, Rilling et al. 1984). Some of the substances used in AgI mixtures are oxidants and may corrode the metal parts of some IN-generating equipment. Solutions may be obtained premixed or can be mixed in the field (Figure 3-7). The AgI must be thoroughly dissolved because undissolved reagent can block flow in the generator, resulting in failures.

Once produced, some AgI aerosols, particularly those containing ammonium iodide, may lose some of their glaciogenic capacity within an hour or so of generation if exposed to direct sunlight. Exposure to sunlight, and ultraviolet (UV) light in particular, may accelerate the deactivation process for some aerosols, while others have shown limited degradation with exposure to sunlight. The degree of photodeactivation, if any, depends upon the formulation and the amount of incident UV solar radiation. In any instance, solutions should be protected from UV light during storage to avoid possible loss of effectiveness.

3.3.2 Dry Ice

Another glaciogenic seeding technique creates cloud ice particles by dispensing dry ice (CO_2) pellets (Figure 3-10) into the supercooled cloud. Dry ice modifies the natural ice formation process by rapidly transforming nearby vapor and cloud droplets into ice (Schaefer 1946, Holroyd et al. 1978, Vonnegut 1981).

Compared with silver iodide complexes, this system has an advantage in that it makes use of a natural

FIGURE 3-7. Acetone-based seeding solutions may be mixed as needed in the field or delivered premixed to field sites. To eliminate evaporation and photochemical reactions, storage of such solutions should be in opaque, air-tight containers. Shown are black plastic carboys that have a 50 L (13 gal) capacity. (Photograph by Bruce Boe, North Dakota Atmospheric Resource Board.)

FIGURE 3-8. Wing rack for burn-in-place cloud seeding flares. Flares shown have a yield of 70 g (0.15 lb), are about 25 cm (10 in.) in length, and have a burn time of 2 min. (Photograph by Bruce Boe, North Dakota Atmospheric Resource Board.)

FIGURE 3-9. Belly rack for the transport and ignition of ejectable cloud seeding flares. The flares, seen three abreast at the front of the rack, are 20 mm (0.79 in.) in diameter; each contains 20 g (0.04 lb) of nucleant. (Photograph by Bruce Boe, North Dakota Atmospheric Resource Board.)

substance (frozen carbon dioxide, CO_2, which sublimes at $-78°C$ ($-108°F$) at 1,000 hPa). However, effective delivery of the CO_2 requires direct injection by aircraft, at altitudes above the freezing level. The CO_2 is also difficult to store, as sublimation (and therefore loss) is continuous. A hail suppression project using this seeding agent was carried out during the 1970s in South Africa. It is not common for dry ice to be the only seeding agent used in a hail suppression project, although it is sometimes used in conjunction with AgI complexes, as is currently the case in North Dakota and Kansas.

3.3.3 Other Ice Nucleants

Certain proteins derived from a naturally-occurring bacterium, *Pseudomonas syringae*, fall within the description of nucleating proteins, because of their ability to induce the formation of ice crystals in seeding applications. This is but one example of the many other substances that have this property. Among these are metaldehyde and 1,5-dihydroxy naphthalene, which have contact freezing temperatures of $-3°C$ ($+27°F$) and $-6°C$ ($+21°F$), respectively. Their efficiency in generating ice crystals is very similar to that of dry ice (Kahan et al. 1995). At present, however, no

hail suppression projects that make operational use of these materials are known.

3.3.4 Hygroscopic Agents

Since the 1950s, hail suppression projects have been making use of AgI complexes as their primary nucleating agent. Nevertheless, the injection of hygroscopic agents that can alter the initial cloud droplet spectra may be an efficient method for treating warm-based cumulus clouds, in which the vertical distance from cloud base to the freezing level can be as much as several kilometers. Appleman (1958) and Ludlam (1958) early on described the concepts involved in seeding with salt particles to prevent hail formation and on a natural hail suppression mechanism, which might explain the lower probability of hail precipitation in storms that are warm-based. Salt was used experimentally in the North Dakota Pilot Project, a combination hail suppression and rainfall enhancement project, in 1972. Hygroscopic seeding for hail suppression was employed on at least one project in the U.S.S.R. during the 1960s (see Lominadze et al. 1973). More recently, following experiments carried out in South Africa in the early 1990s, it has been seen that fresh concepts also are involved and that these have

FIGURE 3-10. Dry ice, in extruded pellet form. The pellets shown, of varying lengths, are about 1.6 cm (0.63 in.) in diameter. Pellets of 0.95 cm (0.37 in.) diameter or slightly smaller are preferred, if available. (Photograph by Bruce Boe, North Dakota Atmospheric Resource Board.)

underlined the potential importance of seeding with hygroscopic agents. Mather strongly recommends the use of hygroscopic agents to combine hail suppression with rain enhancement activities (Mather 1991, Mather and Terblanche 1994).

Hygroscopic agents deliquesce (that is, become liquid by absorbing moisture from the air) at relative humidities significantly less than 100%. Mather (1991) has made use of flares containing primarily potassium perchlorate (Figure 3-11), which when burned produces potassium chloride (KCl) particles of about 1 μm (0.0004 in.) diameter. The two applications are distinct in that the first makes use of large salt particles (requiring payloads of hundreds of kilograms of salt) to create raindrops directly, while the flare technique introduces small particles, only slightly larger than the naturally occurring cloud condensation nuclei (CCN), to enhance the coalescence process in the cloud. The hygroscopic flares weigh only a few kilograms (6 or 7 lb). Although there are many naturally occurring hygroscopic substances, KCl particles have an advantage of only requiring relative humidities on the order of 70 to 80% for deliquescence and readily act efficiently as CCN.

FIGURE 3-11. Hygroscopic flares on wing rack. Each flare contains 1 kg (2.2 lbs) potassium salts and oxidizers. (Photograph by Terry Krauss, Weather Modification, Inc.)

Project planners should bear in mind that the hygroscopic flare method is relatively new and is not yet used as widely as the AgI complexes, but it has shown considerable promise (Cooper et al. 1997, Mather et al. 1996, 1997). A project in southern France is experimenting with hail suppression based on the new hygroscopic flare technique at the time of writing; other experiments are being conducted in Mexico for rain enhancement (Bruintjes et al. 1999).

3.3.5 Quality Control

There are not many manufacturers of seeding agents in the world, but some do offer to prepare seeding chemicals or solutions for weather modification purposes. From the outset, project planners have two options: to prepare their own seeding agents, or to buy them commercially. Whichever option is chosen, the following should be considered:

1. Manufacturers of pyrotechnics should be required to present evidence of their product's nucleating ability, tested in a cloud chamber. Other seeding agents, including those prepared by project personnel, should be tested with the equipment they intend to use before commencing operations.
2. Both active and inactive ingredients of all seeding agents, including flares, should be specified.
3. The quality of all materials must be specified and assured.
4. Any acetone-based AgI solution should be transparent, i.e., colorless to faintly yellow.
5. If flares or rockets are purchased, their warehousing must be carefully planned, as explosive materials are involved.
6. Material Safety Data Sheets (MSDS) should accompany all materials, whether in raw form or ready for use.
7. Dry ice pellets must be of consistent sizes and be transported and stored in containers that minimize sublimation and limit the development of water ice within the dry ice.

3.4 METEOROLOGICAL DATA COLLECTION

Meteorological data collection is necessary in the short-term for forecasting and cloud treatment purposes and in the long-term for project evaluation. Both are essential for program success. Short-term opportunity recognition is the identification of those cloud systems and cloud regions capable of producing damaging hail, and it is immediately followed by the application of the hail prevention seeding. If a storm threatens the target area, the proper time and place for the application of the treatment must be determined. This requires knowledge of the storm location and motion, as well as the local mesoscale and synoptic meteorological conditions. Quality short-term forecasting is thus needed to stay abreast of developing weather, because as new

storms develop, existing storms can persist, intensify, or decay, or change direction and speed.

3.4.1 Forecasting / Nowcasting

"Response time" is the time required to recognize a potential hail threat, notify appropriate responders (pilots, airspace use coordinators, ground-based generator operators, etc.), and initiate seeding. Because thunderstorms can develop very rapidly in unstable atmospheres, and because cost most always precludes having responders always at the ready, day and night, forecasting and nowcasting (very-short-term forecasting in the 0- to 60-min time frame) play a vital role in any hail suppression project. Better forecasts translate to more on-time treatments of potential hail clouds.

3.4.1.1 Government Forecast Information.

The U.S. National Weather Service (NWS), Meteorological Service of Canada, and analogous organizations in other countries provide forecasts about expected storm conditions, and these should be available to the operators of hail suppression programs in these locales. In particular, where official government forecasts of severe weather (large hail, severe winds, tornadoes) or flooding are issued, it is important to be aware of any warnings that apply to the operational area. Policies must be developed for reacting to the issuance of a severe weather advisory or warning in the area. For instance, it may be decided that it is best to suspend cloud treatment operations when certain classes of severe weather have developed, e.g., tornadoes (see also Section 3.6).

3.4.1.2 On-site Operational Forecasts.

Some forecasting should be carried out by the project itself. Only the presence of a skilled forecaster, knowledgeable in the forecasting of thunderstorms and in the local effects influencing them, can ensure the timely direction of operations. Such a forecaster will need meteorological information, to be integrated with the radar and other information available at the operations center, as well as information from personnel in the field, in order to update forecasts on a very short time-scale (nowcasting). Some weather modification operations have found it necessary to operate their own rawinsonde (balloon-borne weather instrument packages that measure temperature, moisture, and winds) equipment in order to obtain the atmospheric temperature and moisture profile and winds aloft, close to the target area, and on time schedules critical to their operations. In the United States, rawinsonde observations are made by the NWS only in the early morning and early evening, while thunderstorm activity begins in the early afternoon. The meteorological information utilized by the forecaster might include surface and upper air observations and also forecast maps produced by the large national or regional numerical forecast centers.

3.4.1.3 Computer-based Radar Storm Tracking.

An example of useful automated radar analysis is the family of programs for the automatic identification and tracking of individual storms in systematically recorded volumetric radar data (Figure 3-12). One such program is TITAN (thunderstorm identification, tracking, analysis, and nowcasting), developed in South Africa and refined at the Research Applications Program of the National Center for Atmospheric Research (NCAR) in Boulder, Colorado (Dixon and Wiener 1993). Storms are identified on each volume scan as volumes enclosed by an envelope composed of a sur-

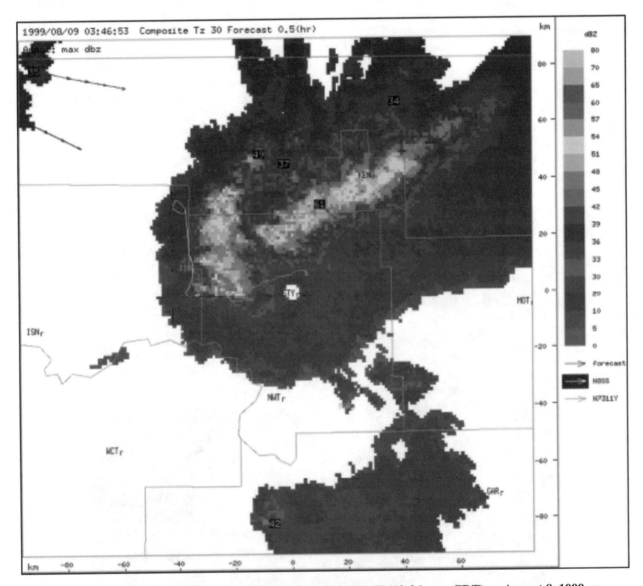

FIGURE 3-12. Composite TITAN radar reflectivity at 03:36 GMT (10:36 p.m. CDT) on August 8, 1999, as recorded by a TITAN-equipped radar. The flight tracks for the preceding 10 min of two cloud seeding aircrafts are also shown. The curving white line near Tioga (TIO) shows the path of a cloud top-seeding aircraft, and a similar line just north of the radar depicts the path of a cloud base-seeding aircraft. The light gray areas between Tioga and Kenmare (KEN) mark the most intense portion of the storm. (Courtesy North Dakota Atmospheric Resource Board.)

face of threshold reflectivity, and a complex algorithm associates a storm cell on one scan with its position on the next.

At each scan, a large number of parameters (height, volume, mass, etc.) are calculated for each storm, and the time history of these constitutes a description of their life cycle. At the discretion of the operator, the storm outlines can be overlaid on the radar display at each scan time. Storm positions in past scans are shown in one color, the present in another color, and forecasts for future storm positions in yet another. The prognostic cell positions are computed from a forecast algorithm, which continues to be improved.

3.4.2 Remote Sensing

It is now possible to obtain useful information about storm development and the near-storm environment in real-time, without having to enter the cloud itself or its immediate vicinity. These techniques, termed "remote sensing," include radar, satellite imagery, and lightning detection networks. The following sections describe some of these techniques.

3.4.2.1 Meteorological Radar.

Aside from the treatment equipment itself, the most important tool for operating a hail suppression operation is the meteorological radar (Figure 3-13). A quality, properly sited radar enables storm monitoring over a large area, in spite of the presence of intervening clouds. The monitoring can provide measurements of vertical development (height of storm tops), intensity (radar reflectivity), location, motion (direction and speed), and, with an experienced operator, the probability of hail development or even the presence of hail. Choices of wavelength and other characteristics of the radar are important. Wavelengths of 3 cm (1.2 in., X-band) or less are not satisfactory for observing mature convective storms, due to the great likelihood of significant attenuation (absorption and scattering) of the radar beam by rain and hail. However, attenuation is negligible at wavelengths of 10 cm (3.9 in., S-band) or greater. This makes it the best choice for convective storm applications, but many projects opt for a 5 cm (2 in., C-band) wavelength, mainly for reasons of cost. The antenna size (diameter) required to achieve a

FIGURE 3-13. Weather radar installation. Shown is a 5 cm (2 in.) wavelength, non-Doppler facility in Stanley, North Dakota. The dish and pedestal are housed within a fiberglass radome that allows operations to continue even during strong winds. The building is air-conditioned to provide comfort to the electronics and to project staff. An auxiliary power unit (APU, in foreground) provides power when commercial service fails. (Photograph by Bruce Boe, North Dakota Atmospheric Resource Board.)

given beamwidth (typically between 1 and 2 degrees) at 10 cm (3.9 in.) wavelength is twice that at 5 cm (2 in.). For example, a 5.4 cm (2.1 in.) wavelength (C band) circular-parabolic radar antenna must be 2.4 m (7.9 ft) diameter to produce a nominal 1.6 degree beamwidth, whereas the same beamwidth at S-band wavelengths requires an antenna of 4.9 m (16 ft) diameter. The cost is increased by an even greater multiple. To ensure adequate coverage of the target and adjacent areas and to minimize the "ground clutter" detected by the radar, the selection of radar hardware and the siting of it should be made with the advice of an expert in radar meteorology.

It would be possible to carry out a hail suppression operation in many locations using only the publicly available radar information such as might be obtained from the NWS (National Oceanic and Atmospheric Administration [NOAA]). However, it is preferred, especially in projects utilizing faster delivery systems such as aircraft or rockets, for each project to have an independent radar with full control, to ensure that the very latest information is available and updated on a time scale of a few minutes.

A project radar can be operated manually to obtain the information needed to guide cloud treatment ac-

tions, but manual operation does not lend itself well to the fullest documentation of storm development and morphology. Modern digitized radars allow systematic recording of storm data on tapes, disk, and CD-ROM, so that operations can be documented, reviewed, and, if desired, analyzed. Data obtained with a well-calibrated digital weather radar also comprise an important resource for storm research and for program evaluation. Digitized radars can be operated under computer software control, and there exist various software techniques for performing sophisticated analyses of the data, either in real-time or from the recorded data.

3.4.2.2 Satellite Data.

Satellite information can be of considerable use in conducting hail suppression operations, mostly for forecasting thunderstorm activity. The best sources for this information are geostationary satellites, whose orbits are just high enough that their speed matches that of the rotating earth below. Thus, these geostationary satellites always remain above the same spot and are able to continuously provide imagery in visible, infrared, and "water vapor" wavelengths (Figure 3-14). The NOAA Geostationary Operational Environmental

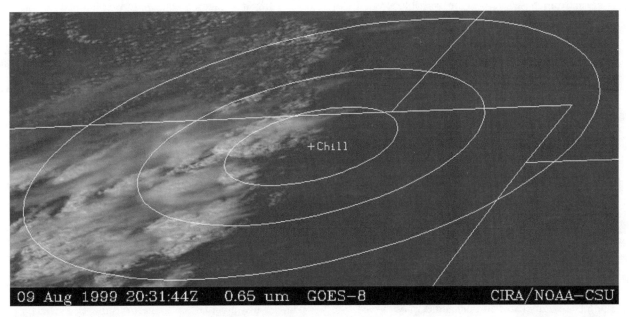

09 Aug 1999 20:31:44Z   0.65 um   GOES-8   CIRA/NOAA-CSU

FIGURE 3-14. Examples of GOES satellite imagery. Visible-wavelength imagery of 1 km (0.6 mi) resolution is shown for the vicinity of the CSU-CHILL research Doppler weather radar (lower). It reveals numerous convective cloud development in the same mountains of Colorado during the mid-afternoon. An infrared image at the same time (upper left) shows the same general pattern, though at lesser (4 km, 2.5 mi) resolution. A water vapor image of the central U.S. (upper right) shows a dry slot over the mid-Mississippi Valley, with numerous small thunderstorms in the southeast showing up as white dots. (Imagery courtesy of the Cooperative Institute for Research in the Atmosphere, Colorado State University, Fort Collins.)

Satellites (GOES) provide excellent coverage of the Americas. Analogous coverage is provided of Europe, Africa, and Asia by METEOSAT satellites, operated by the European Community.

Visual-wavelength imagery is generally available at resolutions as small as 1 km, (0.62 mi) although only during daylight hours. The newest GOES satellites now routinely provide visual wavelength imagery of sufficient quality that cloud structure can be resolved on the flanks of mature thunderstorms, in some cases even the larger individual cloud turrets. Infrared imagery resolution is somewhat coarser, generally 4 to 8 km (2.5 to 5 mi) per pixel (picture element). This means that less detail can be resolved in the infrared, but, unlike the visual wavelength imagery, it is available throughout the night as well. Infrared satellite imagery provides a measure of the temperature of the cloud top, which can be combined with appropriate sounding data to derive cloud top height. Water vapor imagery is derived from an infrared band that senses the amount of water vapor of the top third of the troposphere. Moist areas show up as white, dry areas as black. This imagery is very useful in detecting higher clouds, such as those associated with jet stream motions, and consequently is helpful with forecasting.

Data from these satellites are available at many government and university sites on the Internet. In addition, imagery can be received directly by means of on-site receiving equipment. Use of satellite images in conjunction with radar and lightning data presents a powerful technique for following developments in convective systems organized on a large scale.

3.4.2.3 Lightning Location.

Automatic near-real-time systems for display of the locations and polarities of cloud-to-ground lightning now cover the entire United States (Orville 1994), as well as the United Kingdom (Lee 1986). These systems can be of great value for they identify, in real-time, active convective storm systems that are generating cloud-to-ground lightning (Figure 3-15). In addition, the character of the observed lightning flashes themselves can provide indications regarding storm life cycle or intensity (MacGorman and Rust 1998), or severity (MacGorman and Burgess 1994). The lightning information supplements radar and satellite data, positively identifying electrically active clouds even at great ranges from project radars.

3.4.3 Visual Cloud Appearance

Visual cloud appearance is often the primary criterion upon which treatment decisions are based. It allows treatment decision-makers to assess the growth stage of the cells of interest, and allows pilots to gauge the potential risks associated with treating them. While radar echoes provide a measure of the strength of mature cells, it is the nonechoing developing cells that must be targeted (Figure 3-16). During passage of a storm, cloud features and motions can be very misleading to untrained persons, as can be appreciated by interviewing eyewitnesses after a hailstorm. Even the direction of storm movement can be erroneously perceived.

Experienced weather modification pilots and well-positioned ground-based observers can provide helpful real-time visual storm reports that describe what a storm is doing at that moment. Project radars typically detect only clouds that have begun to produce precipitation embryos. This means that an observer can see and report new cloud development before it produces radar echoes, sometimes minutes before the radar operator has any chance of seeing it on the radar. These minutes can be critically important, for it is these growing, supercooled, ice-free cloud turrets that must be targeted. It is, therefore, very important not to rely solely on radar.

3.4.4 In Situ Measurements

In situ measurements are those obtained by entering the place to be measured; they are helpful in quantifying real-time cloud conditions, including updraft, ice content, liquid water content, and temperature. In hail suppression programs employing aircraft for delivery of the seeding agents at cloud top, in situ measurement can be accomplished by instrumenting the seeding aircraft. The flight crew will then know as they penetrate the potential target turrets how suitable for treatment such turrets are. The additional expense is not excessive, and the information gained is most helpful. In addition, the data will be available for post-analysis if efforts are made to refine and improve treatment procedures, which is strongly encouraged.

It should be noted that all instruments must be calibrated. This can be achieved either by flight past calibrated instruments on a tower or by flight in close proximity to other aircraft that have well-calibrated instrumentation. Some instruments, such as those for measuring cloud water content, can be accurately calibrated only by operation of the instrument in a carefully controlled environment, such as a wind tunnel, in which varying concentrations of liquid water are established. Such calibrations are generally done in a laboratory environment rather than in the field.

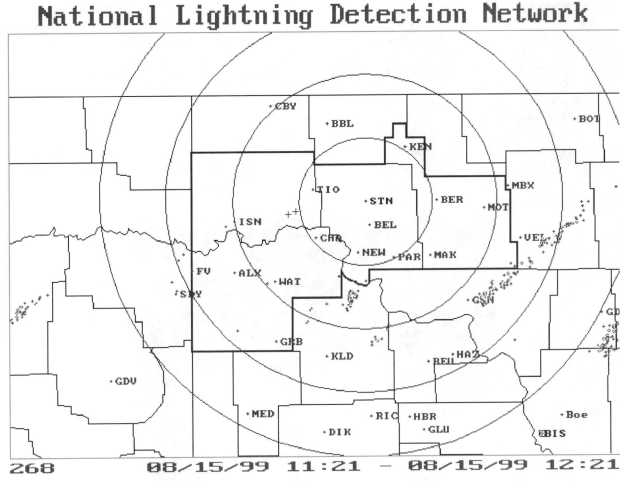

FIGURE 3-15. Map showing the location of cloud-to-ground lightning flashes during a 1-h period, relative to a hail suppression project radar (STN). Radar range circles are in 46 km (25 NM) intervals. The National Lightning Detection Network data are obtained in near-real-time via satellite downlink. Software allows archival and playback of data, which can be exported to a spreadsheet.

3.4.4.1 Updraft.

Growing turrets, particularly those just becoming supercooled, are those most suitable for glaciogenic treatment. The strength of the turret updraft is a good indicator of growth rate and hence cloud vigor. Aircraft penetrations should always be flown with the aircraft in level flight; therefore, updrafts and downdrafts will register as altitude changes. Because one of the criteria for treatment should be the existence of an updraft (perhaps of some threshold speed), it is desirable to have real-time knowledge about the updraft character. Thus, instrumentation that will reliably provide near-instantaneous data is desirable. These instruments range, in order of simplicity, from the vertical velocity indicator (VVI) and the instantaneous vertical velocity indicator (IVVI), to

GPS, to flight management systems, and even to inertial navigation systems (INS). An experienced radar meteorologist can deduce approximate updraft strengths from Doppler radar velocity data by identifying areas of low-level convergence, but this technique works only with echoing (mature) storm cells, not young cumulus congestus cloud desired for treatment. Thus, operators must resist the temptation to ascribe such updraft observations to suitability for seeding.

3.4.4.2 Cloud ice content.

It is important to know, at least in a relative sense, how much ice is present within the cloud under consideration for treatment. Without instrumentation designed for this specific task, the flight crew is reduced to visual observation while in-cloud, perhaps augmented by the audible "click" of graupel or hail as it impacts the windshield. Instrumentation tailored to this purpose ranges from simple ice particle counters (IPC) to sophisticated optical array instruments, such as the Particle Measuring Systems (PMS) one- and two-dimensional cloud and precipitation probes (Knollenberg 1970, 1976; Figure 3-17), which can provide hydrometeor spectra data (numbers and sizes of hydrometeors). The latter cannot be utilized without implementation of on-board computers to process, display, and archive the data. Cloud particle imaging (CPI) equipment (Figure 3-18) is now available that offers remarkably clear images of hydrometeors (e.g., Korolev et al. 1999). Another relatively new instrument, the high-volume precipitation spectrometer (HVPS), has greatly improved the statistics of precipitation particle sampling (Lawson et al. 2000).

3.4.4.3 Cloud Water Content.

For glaciogenic seeding to be effective, the subject cloud must contain supercooled liquid water (see

FIGURE 3-16. The view forward from an aircraft seeding at cloud top. Note the crisp, white cloud top, indicative of growth (updraft) and suggestive of a lack of ice. This cloud would not yet be detectable by a ground-based radar. (Photograph by Bruce Boe, North Dakota Atmospheric Resource Board.)

FIGURE 3-17. PMS two-dimensional-C optical array probe, carried beneath an aircraft wing. A laser beam is shone between the two "arms" of the probe, upon an array of photodiodes. The number and sequence of shadowed diodes determine the size and shape of each passing particle. (Courtesy Andrew Detwiler, South Dakota School of Mines and Technology.)

Section 3.4.4.4). For quantification of the cloud water content, several options are available. Without instrumentation, the flight crew usually can gain a reasonable qualitative feel for cloud water content just by observing the rate of ice accretion on the leading edges of the airframe and windshield. For measurements that are more quantitative, instrumentation ranges from simple hot-wire sensors such as the Johnson-Williams (J-W) cloud water probe, to the CSIRO-King probe, to the PMS forward-scattering spectrometer probe (FSSP; Figure 3-19), to the cloud drop spectrometer (CPS). The PMS FSSP and the CPS record the cloud droplet spectrum itself, which is very useful information when hygroscopic seeding is being conducted or considered.

3.4.4.4 Temperature.

Temperature must be known to determine the degree of supercooling. Simple outside air sensors are effective when the aircraft is flying in clear air. Accurate sensing of temperature during flight within-cloud becomes more difficult. This is because cloud droplets first wet, then evaporate from the sensing surface, resulting in evaporative cooling. Temperature sensors that use an aerodynamically induced reverse flow effectively avoid this problem by sampling air taken in from an aperture located on the instrument's downwind side. The "reversing" of the flow is possible only for the air itself; the cloud droplets are too massive and cannot change direction fast enough to make the turn into the sensor, and they

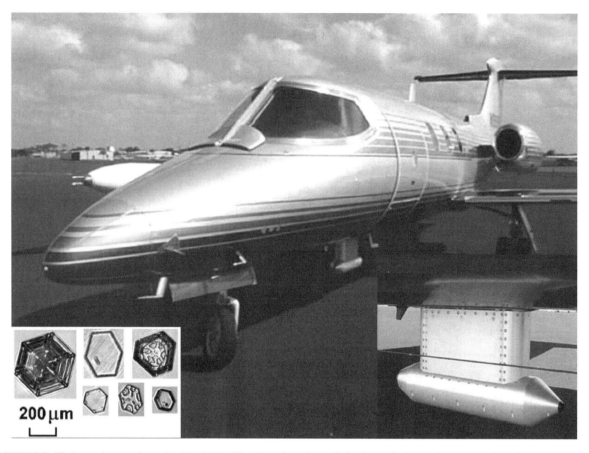

FIGURE 3-18. Lear jet equipped with CPI affixed to the aircraft belly and shown in lower right inset. Sample images are shown in lower left (Korolev et al. 1999). (Photo of jet and CPI courtesy Paul Lawson, Stratton Park Engineering Company, Inc.)

therefore pass on beyond the instrument. Such instruments provide for improved in-cloud temperature measurement, without the effects of evaporative cooling. The locations of these instruments on the airframe also are extremely important. Care needs to be taken to avoid placement in locations where changes in aircraft attitude can result in erroneous measurements.

3.4.5 Numerical Cloud Modeling

The ever-increasing speed and computational power of computers allows the physical processes within the atmosphere to be modeled with equations, beginning with prescribed conditions matching some (known) initial atmospheric state. These iterative cloud models can be in one, two, or three dimensions, and may be time-dependent or be steady-state models. The complexity of a model is determined by the trade-offs between time and space resolution and degree of detail used in representing the dynamical and microphysical

processes of formation and development of cloud, rain, and hail (Orville 1996).

In many operational applications, one-dimensional models serve as a refinement of the classical techniques for analyzing the atmospheric soundings (obtained by weather balloon instruments) to predict things such as vertical storm development and maximum probable hailstone size. Multidimensional models are currently too complex for use in other than research projects; however, with advancing computer technology, this should soon change.

3.4.5.1 Predictive modeling.

Some existing cloud models can be initialized with local conditions (a sounding) and run quickly enough that the output can be useful in field programs. In two recent North Dakota field programs (Boe et al. 1992, Boe 1994), a two-dimensional, time-dependent (2D-TD) cloud model was initialized daily with the atmospheric data collected by the balloon-borne sounding made lo-

FIGURE 3-19. PMS FSSP for the measurement of cloud droplet spectra. The operating principle is analogous to the two-dimensional-C probe, but with different optics. (Courtesy Andrew Detwiler, South Dakota School of Mines and Technology.)

cally by the NWS at 7:00 a.m. local time. The model was then "run" on a Cray supercomputer at the NCAR in Boulder, Colorado, and the resulting predictions for the day's weather were downloaded in time for the daily 11:00 a.m. local time project weather briefing (Kopp and Orville 1994). The scale of cloud development (if any) and the timing of the onset of convection were both predicted and presented to the researchers gathered for the morning weather briefing as part of the routine forecast. Figure 3-20 shows one aspect of model performance in a predictive mode. Computers also are being used to model airflow in complex terrain (e.g., Stone 1998). Such models can be helpful in planning the placement of ground-based aerosol generators.

3.4.5.2 Diagnostic modeling.

Cloud-scale computer models also can be employed post-hoc to examine the evolution of observed cloud systems. Such models employ knowledge of various cloud processes, both dynamic and microphysical, to simulate cloud development. A good indicator of the validity (accuracy) of such cloud models is how well they replicate the actual cloud systems observed, given the best knowledge of initial environmental conditions. Cloud models are helpful in evaluating observed cloud behaviors, including the effects of treatment, and often yield insight into processes not previously understood (e.g., Farley and Orville 1999). However, they are not useful in real-time decision-making.

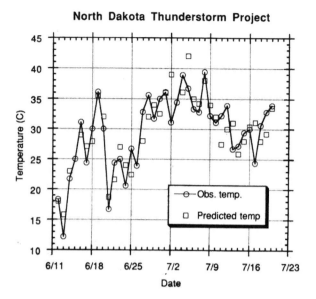

North Dakota Thunderstorm Project

FIGURE 3-20. The observed maximum temperature for each day of the North Dakota Thunderstorm Project, and the corresponding model-predicted temperature. (From Kopp and Orville 1994, copyright American Meteorological Society, used by permission.)

Because no two clouds are exactly the same, it is impossible to find two identical clouds, seed one, and leave the other untreated as a control. However, the effects of seeding can be examined by modeling two clouds that begin with identical initial conditions, and simulate seeding in one but not the other. Any differences in behavior can then be attributed to the seeding. In fact, this option is very attractive, as the timing (relative to the life cycle of the subject cloud), locations, and amounts of seeding can be varied in a succession of model runs to better understand the importance and effects of targeting. Of course, caution must be used in drawing conclusions based on the modeling of only single case-days. Output from such a diagnostic model run is shown in Figure 3-21.

3.5 SELECTION AND SITING OF EQUIPMENT

The optimum siting of equipment is critically important to the success of any program. Whether aircraft, ground-based generators, or rockets, the deployment of these facilities relative to the target area must be carefully planned. Response time and expense must be minimized. Safety, ease of access, and convenience should be maximized for project personnel.

There are many types of seeding equipment available, and usually a large number of locations where such equipment might be based. Once the seeding methodology has been determined (see Sections 3.2 and 3.3), the local storm climatology, based on radar data if records are available, and the available funding usually determine what equipment is deployed.

For example, if three seeding aircraft can effectively treat as many thunderstorms as usually develop at one time, and funds are available to deploy three aircraft, the project should plan to deploy three aircraft. It is likely that an unusual number of storms will occasionally develop. When this occurs, three aircraft might not be enough. Nevertheless, four aircraft would not be deployed (in this example), because the additional cost does not warrant deployment of facilities that would be largely unused. This consideration drives the deployment of all facilities, whether they are aircraft, ground generators, or rocket launchers.

3.5.1 Aircraft Considerations

Aircraft employed in cloud treatment operations in or around thunderstorms should possess certain minimum performance characteristics, to ensure operational and flight safety requirements. Light, single-engine aircraft are used on some projects primarily due to their reduced operating costs. However, many projects operate twin-engine aircraft in the belief that if one engine malfunctions, the aircraft can return under power to a nearby airport. Unpowered, forced landings in the dark, in the vicinity of thunderstorms, are dangerous.

For direct injection at the tops of growing super-cooled clouds, some operations employ turboprop or small twin-engine jet aircraft. Piston-driven aircraft can perform satisfactorily at cloud top; but they are slower and have reduced rates-of-climb, requiring a much greater initial response time. Although pressurization of the aircraft cabin is not required if oxygen is available to the top-seeding flight crews, it is desirable. Oxygen masks are not needed, and the greater comfort reduces crew stress.

Aircraft types proposed for cloud base and penetration seeding of cumulus turrets should be examined carefully, and their relative performance specifications should be weighed in consideration of the operational design. Program sponsors may choose to require demonstrations of aircraft performance as a condition of the contract award, to ensure that the aircraft are capable of performing as bid. Sponsors also may wish to review the aircraft maintenance history.

When aircraft are to be used for hail suppression seeding operations, close coordination is necessary

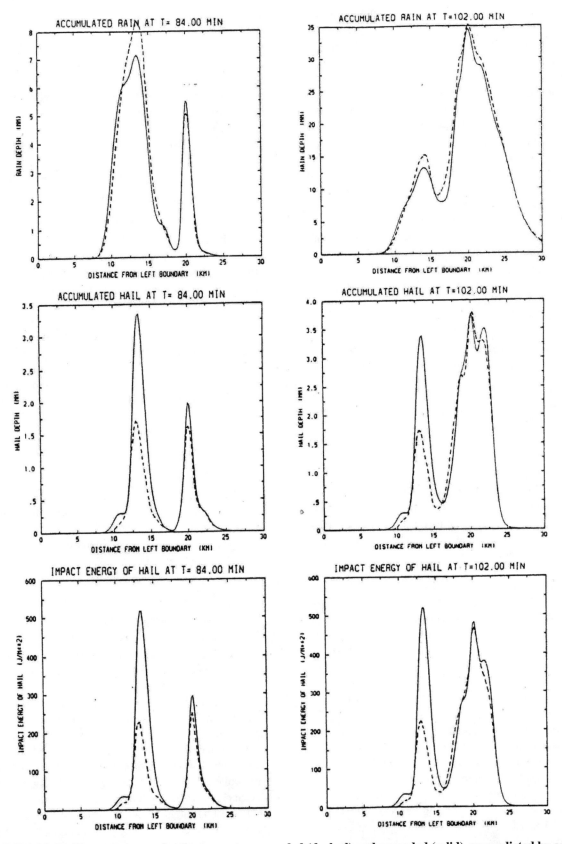

FIGURE 3-21. Hail impact energy for the same storm, seeded (dashed) and unseeded (solid), as predicted by an analytical cloud model. (From Farley et al. 1996, copyright American Meteorological Society, used by permission.)

between those directing operations and the flight crews. Since flight crews benefit from being able to view the project radar and visit with project forecasters before undertaking operations, it is desirable to base at least a portion of the project aircraft fleet at an airport near the project radar. However, advancements in communications technology and the ready access to weather information via the Internet may make it possible for some project personnel to monitor the project radar imagery from off-site. If pilots can see the radar depiction of the developing storm complex before they are dispatched to intercept it, they will have a better grasp of the situation and will be more certain in their actions. When target areas are large, aircraft should be based at several airports within or near the target. Pilots who do not have access to real-time radar information should be provided quality preflight weather briefings by the project meteorologist. Of course, all airports must have ready supplies of fuel, adequate runways, and navigation aids. Maintenance services are not strictly required but are desirable.

3.5.2 Placement of Ground Generators

The number of ground-based ice nuclei generators required for hail suppression operations varies depending upon target size, topography, and the nature of the prevailing wind regimes. The nuclei produced by the generators can reach the growing cloud systems in two ways: (1) by convective currents (thermals) produced by surface heating, or (2) by lift provided by flow over the topography in cases where mountains are present and local winds are favorable. Project planners should study the project area climatology to ascertain which of these, if either, is likely to be the primary vertical transport mechanism. The time required for transport from the surface to cloud base altitude will vary considerably depending upon the vigor of the convection, from as little as 15 min to as much as several hours. Therefore, generators must be activated in advance of anticipated storm development. This requirement underscores the importance of accurate forecasting. Targeting a fixed area with ground-based generators requires a portion of the generator network to be deployed upwind of the target, which means placement of generators outside the target area on several sides, to account for all common upwind flow regimes. Surface winds tend to converge into areas where new convective development is occurring, so operating generators 15 to 25 km (9.3 to 15.5 mi) in advance of vigorous storms can be useful (Dessens 1998, Figure 3-1).

Although it has been shown that transport of surface-based seeding agent releases to convective cloud base can be accomplished (Heimbach and Stone 1984), this should not be taken for granted. Seeding must be done over large areas because releases drift with the environmental winds. Individual clouds are generally not targeted. Ground-based seeding is sometimes chosen because the deployment of aircraft is not required. However, significantly more seeding agent may be required, as seeding most often must start earlier, continue longer, and be conducted over a larger area.

3.5.3 Placement of Rocket Launchers or Artillery

The greatest advantage of treatment via rocket or artillery is the extremely small response time if positioned within range of the target cloud(s). Sites are normally fixed (nonmobile); so given the relatively short ranges of most rockets and artillery (up to 12 km, or 7.5 mi), many sites will be required to ensure coverage of most target areas. Launch sites can be determined by drawing circles of the same scale as the maximum range of the rockets or artillery and moving the circles around on a map of the proposed target area until the desired coverage is achieved.

3.6 LEGAL ISSUES

The legal aspects of cloud seeding relate to regulation by law and to the possibility of litigation by persons who feel they have been or would be harmed by such activities. Environmental concerns are described in Section 3.7. The legal considerations for hail suppression are essentially those set forth for precipitation augmentation operations in ASCE Manual No. 81 (Kahan et al. 1995).

3.6.1 Potential for Litigation

Project sponsors must address the potential for lawsuits, even if no unusual weather occurs during the project period. Sponsors should request the contractors who provide weather modification services to attempt to obtain liability insurance against the effects of operations. This special insurance is termed "consequential loss insurance" and is not normally a part of ordinary liability insurance. Since severe storms are being dealt with in hail suppression operations, concerns relating to public safety and well-being may be heightened. Perhaps the greatest concern is that, for whatever reason, seeding might be perceived to produce more or larger hail or excessive rains. This scenario is the primary reason that project sponsors and operators should carry consequential loss liability insurance covering the effects of weather modification. Since the burden of proof rests with the plaintiff, proving such a claim is unlikely.

Another concern relates to the treatment of tornadic or funnel-bearing clouds. The regulatory entity in the State of North Dakota has taken the position that such storms are not to be seeded (Boe et al. 1998). If a seeded storm should develop a tornado, treatment is to cease immediately and not resume until at least 30 min have elapsed after the dissipation of the tornado or funnel. This decision was not made lightly, as tornadic storms are usually among the most severe and often produce significant damaging hail. The decision not to seed tornadic storms arose from the concern that litigation could result if a treated tornadic storm was to result in loss of life or major property damage.

Conversely, tornadic storms in the State of Kansas are frequently treated for hail suppression because treatment is not known to exacerbate the tornadic condition. Too much hail falls from tornadic storms to cease seeding whenever a funnel appears. In this case, the risk of litigation is outweighed by the risk of major hail damage (Smith 1999).

3.6.2 Regulation

Most locales have some regulatory agency responsible for the enforcement of laws or ordinances relating to the conduct of cloud seeding operations. In the United States, Public Law 92-205 requires that all cloud seeding operations be reported to the NOAA. Under this law, initial reports must be submitted that describe the geographical location of operations, the proposed dates of operations, and the seeding agent(s) to be used. A file number is assigned to each project in the United States. Upon conclusion of the project, a final report must be submitted that summarizes the dates of operations, as well as the quantities of seeding agent(s) dispensed. Interim reports may be required. Failure to file these reports within 45 days of the conclusion of the project is punishable by a fine up to $10,000.

Many states have promulgated laws, rules, and regulations that establish the process through which proposed projects must be approved. Usually these require the person(s) who will conduct the operations to become licensed in that state. Licensure considers the credentials, character, and experience of the project manager(s), but not specific projects. Typically, a permit also must be obtained for each specific project area. Thus, both a license and a permit are usually required before operations can be conducted. It is essential that projects comply with local, state/provincial, and federal laws, so project designers must identify all relevant laws and regulations early on.

3.7 ENVIRONMENTAL CONCERNS

It is incumbent upon the project designers to fully address the potential impacts of seeding operations upon the environment. In some locations, such assessment may be required by local, state, or federal law; in others, none may be required. Even when the latter is the case, the type of seeding agent to be employed, and estimates of the quantities to be used, have served as the basis for a simple environmental assessment. Although even the most active programs disperse seeding agents at rates too low to be readily detected by conventional analytical chemistry equipment, this exercise is worthwhile because it will be reassuring to area residents. Only specialized facilities employing the most sensitive analytical chemistry techniques are readily capable of detecting seeding material on a regular basis.

The two most frequent environmental quality issues are (1) the potential for undesired effects to result from seeding, and (2) the possible toxicity of the seeding agent(s). The first of these relates to concerns about excessive or insufficient precipitation. The latter arises because silver iodide complexes, the most common of the glaciogenic agents, are sometimes dispensed over large areas for many seasons and consequently raise questions about potential environmental impact (Kahan et al. 1995). Neither issue has ever been found to pose a problem, however.

3.7.1 Redistribution of Precipitation

Though most current evidence supports slight net increases in precipitation both within and downwind of hail suppression project areas, plans should be made to collect data sufficient to evaluate this aspect of the program. Section 5, which presents some evaluation concepts, offers further guidance in this regard.

3.7.2 Seeding Agent Safety

In the last several decades, numerous projects have conducted environmental assessments or prepared environmental impact statements without any adverse findings. Still, each project may find it helpful to assess the possible short-term and long-term (cumulative) effects of additional silver and/or iodine in the air, water, and soil. In addition, other materials are now being used with increasing frequency in pyrotechnic compositions and seeding solution formulations. Also, trace metal impurities are associated with major components of the aerosols produced by individual combusted pyrotechnic compositions or solution formulations. Thus, the compositions of solution formulations should be reviewed from an environmental impact basis, with respect to the threshold toxicity limit and any local "hazardous materials" restrictions, etc.

The proper procedures for the handling and storage of hazardous materials of all kinds are thoroughly prescribed by local, state, and federal laws, and are the responsibility of the contractor or agencies actually in charge of the operations. Most cloud seeding agents are not considered hazardous. Exceptions include those that are readily combustible, such as pyrotechnic devices and acetone solutions (see Section 4.5).

Hail suppression programs are generally designed to be carried out over a period of several or many years, and the quantity of seeding agent dispersed can be significant. To give an example, in southern France an ongoing hail suppression project covers a target area of 40,000 km^2 (15,400 mi^2) and uses thousands of kilograms of AgI annually (Dessens 1998). This makes it recommendable, especially for long-term projects, to introduce some means of effective environmental evaluation. This provision is more a measure for social and client reassurance than anything else. Klein (1978) has shown that the quantities of AgI released in projects such as this scarcely increase or contribute significantly to the concentrations of atmospheric Ag caused by other human activity. Many hundreds of years would need to elapse before any possible environmental impact could be detectable from this cause.

Many research programs have included investigation of the potential for environmental impacts from seeding agents. These studies began in the 1970s and mostly concluded in the 1980s. The results from these studies concluded that "the major environmental concerns about nucleating agents (effects on plant growth, game animals, fish, etc.) appear to represent negligible environmental hazards" (Klein 1978).

Analytical techniques exist for the detection of silver and other trace metals from air filters, precipitation, and soil sediment samples at parts per billion concentration levels [$<1.0 \times 10^{-12}$ g ($<0.4 \times 10^{-14}$ oz) AgI ml^{-1}] or lower (Warburton et al. 1982). If the possibility for residual effects of seeding agents becomes an issue, well-established procedures are available for addressing it.

4.0 OPERATION OF HAIL SUPPRESSION PROJECTS

Those conducting hail suppression operations must be vigilant, and always prepared to react when potential hailstorms begin to develop. In this section, some guidelines are provided for operational decision-making and project infrastructure. In addition, guid-

ance is offered regarding communications, personnel, safety, public information, and daily routine.

4.1 THE OPERATIONS MANUAL

The operations manual should furnish clear definitions of all terms used in conducting operations, describe the types and functions of all information-gathering systems employed and all actions to be undertaken, define all phases of operations, and state criteria for initiation and termination of each phase. Procedures for recording and reporting operational information should be given, particularly to comply with any relevant weather modification laws. The circumstances under which seeding will be suspended also should be clearly defined, so that prompt decisions can be made when circumstances warrant. Sections 4.2 through 4.6 describe other issues that should be addressed in the operations manual.

New program directors are advised to obtain copies of manuals used on other projects and use them as guidance to form and content. Although they will not be specific to the new project, the manuals will provide an idea of the necessary organizational infrastructure. Commercial cloud seeding firms can usually provide samples of manuals used on their other projects; this is a great first step in creating a new, project-specific manual.

4.2 PERSONNEL REQUIREMENTS

Many of the duties required of operational hail suppression personnel require special knowledge and skills. National, state, and local laws should be consulted to determine whether personnel in control of operations are required to be certified. The state in which operations are planned should be contacted. Some of the specific skills required of various project personnel are discussed in this section. The WMA has established processes for the certification of weather modification operators and managers (WMA 2000).

4.2.1 Meteorological Staff

At least one meteorologist with experience in forecasting of convective weather should be part of every operation. This ensures an adequate understanding and prompt interpretation of all meteorological information flowing into the radar-equipped operations center, from which operations are directed. An experienced project meteorologist can produce forecasts tailored to the local situation and needs and might also be a radar operator and/or operations director. Larger projects, particularly those intending to conduct operations 24 h each day, 7 d each week, must retain sufficient meteorological staff to support around-the-clock operations. Meteorological duties will include forecasting and nowcasting, radar operation, data acquisition, and, often, the direction of seeding operations. How these tasks are assigned is project dependent.

If project staffing provides for more than one meteorologist and treatment is done by aircraft, it is strongly recommended that any project meteorologist not directing operations occasionally fly onboard seeding aircraft. This experience provides the meteorologist with first-hand knowledge of flight conditions aloft and an understanding of the pilot's perspective of storm development.

4.2.2 Cloud Treatment Pilots

Hail suppression projects using aircraft for cloud treatment must be concerned that licensed, IFR-rated pilots be well trained and have experience flying in close proximity to thunderstorms. This can be done safely only if the pilot has the knowledge and experience to avoid dangerous situations.

Projects choosing to deliver seeding agents by aircraft ensure the most certain, direct, and economical use of seeding agents. However, because the pilots are such a critical link in the delivery, every effort must be made to ensure that they are well trained for the task. The pilots hired either must have experience seeding thunderstorms, or they must receive adequate pre-project training, preferably the former. Arrangements can sometimes be made for new pilots to fly with experienced pilots on other projects, or they could be sent to the University of North Dakota for formal instruction in weather modification flight.

4.2.3 Direction of Operations

Operations are usually directed from the project radar by a person experienced with hail suppression and radar operations. This individual, the operations director, may or may not have responsibility for operation of the radar itself, which may be operated by another field meteorologist. At the radar, information important to decision-making must be available to the field operations director. Such decisions typically concern the launching of aircraft or rockets, ignition of ground-based silver iodide generators, the termination of operations, and so forth. Communications are of paramount importance when running hail prevention operations. Radio contact must be maintained between aircraft and the radar center. Cellular telephones and/or pagers can be useful in maintaining

contact with personnel in the field. Communications between the operations center and ground-based generator and/or rocket-launching sites usually can be accomplished by similar means. The operations director is normally the individual accorded the responsibility for suspension of seeding and, often, for the assignment of daily operational status to project personnel. Typical status categories include "weather watch," when no workable clouds are forecast in the short-term; "standby," when suitable clouds are forecast to develop some time in the next few hours; and "alert," when development is either occurring or thought imminent.

4.2.4 Support Personnel

All types of hail suppression operations require personnel to maintain and operate equipment or make observations. Technical requirements are high for persons maintaining radar or other electronic and computing equipment; an engineer or skilled technician may be required. Less-skilled persons can be employed for many other tasks, such as maintaining ground-based silver iodide generators and passive hail sensors (hail pads). Others will be needed to deliver seeding agents to field sites, activate and deactivate ground-based generators, and complete and track project paperwork.

Training needed varies for persons employed to load and ready rocket launchers, depending upon the type of rocket-based seeding system used. Those preparing launchers controlled from the operations center require less skill than those who receive launching instructions verbally, as the latter must load, aim, and fire the rockets, while the former is targeted and fired by computer control.

4.3 OPERATIONAL DECISION-MAKING.

Close coordination is required between the operations director, who determines when to start and stop seeding, and the seeders, who actually dispense the seeding agent. While the field meteorologists, led by the operations director, normally work from a project radar that serves as the operations center, the seeders may be waiting at airports or within their aircraft, or at remote generator or rocket launching sites. The relationships among the various project personnel are illustrated in Figure 4-1.

4.3.1 Chronology

Most projects begin each operational period, usually defined as a calendar day, with a weather briefing,

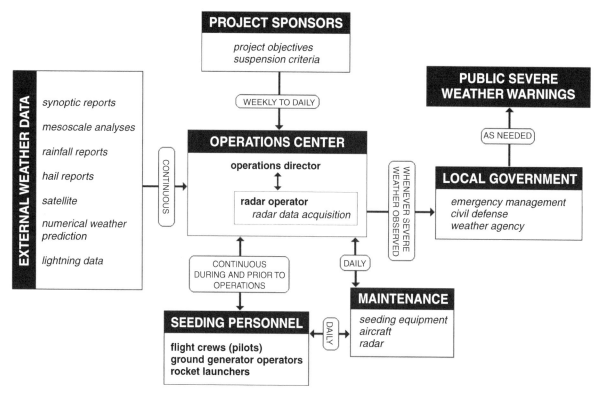

FIGURE 4-1. The interrelationships of project personnel on a typical hail suppression project. The decisions to begin and end operations are made by the operations director at the project radar, in accordance with project objectives and suspension criteria provided by the project sponsors.

wherein project staff are informed by the forecaster of the likelihood of clouds that meet the criteria for seeding as set forth in the operations manual. If such conditions already exist at the time of the scheduled briefing, operations will already be underway. However, the forecast is still given so that the personnel will have some idea as to how long they are expected to continue operations. After the weather briefing (if operations are not underway), duty assignments for the day are made by the operations director.

The balance of the operational period is then spent waiting for suitable clouds, or for operations to be called "down" for the day. Because forecasting is an imprecise science, most operational programs, especially those that conduct operations around the clock, never formally excuse operational personnel from their assigned duties for fear of missing unforecast opportunities. Projects anticipating frequent long periods (> 12 h) of continuous operations should consider retaining sufficient staff for multiple shifts to ensure that all on duty are adequately rested and alert.

Once operations begin, staff are engaged accordingly. If no suitable clouds develop, the operations di-

rector, after consulting with the forecaster, can at some point downgrade the operational status so that personnel are less restricted. For example, if cloud development is initially thought likely, project personnel should be placed on "alert" and thus be compelled to remain at or near their operations stations. If at a later time the forecaster determines that cloud development is no longer likely, personnel should be given a status of "standby" or less. The "lower" status allows them the flexibility to do other things, such as paperwork, simple maintenance chores, etc. The whole cycle repeats the following day.

Invariably, some project days will clearly lack the degree of atmospheric instability necessary for the development of strong thunderstorms. When such days are identified, complex maintenance and other tasks that require personnel to be away from their duty stations can be accomplished.

4.3.2 Opportunity Recognition

Simply releasing seeding agents when thunderstorms are present does not ensure that the seeding agent will reach the targeted cloud turrets and produce the desired result. This statement applies to all treatment methods, unless the seeding agent is released directly in the location of interest. Otherwise, some assumptions must be made. In any case, it is useful for those directing operations to know, with as much certainty as possible, which clouds have been seeded. The degree of uncertainty varies considerably depending upon the treatment method.

4.3.2.1 Aircraft flight crews.

When using aircraft for hail suppression seeding operations, close coordination is necessary between the radar operator/operations director and the flight crews. Flight crews and meteorologists alike must remember that the seedable cloud candidate is not the mature storm, but rather the developing cloud turrets on the flank of the mature storm, or if no mature storm has yet developed, the most vigorous of the developing towering cumulus in the cloud field. The preferred candidates are largely ice-free but supercooled, so cloud top temperatures between −5°C and −10°C (+23°F and +14°F) are most desirable when seeding is done by direct injection at or near cloud top altitude. If treatment is done from cloud base, release of ice nucleating aerosols can begin in the updrafts beneath even smaller (younger) clouds, because several minutes will be required for the aerosol to be transported aloft to supercooled cloud.

The exact positioning of the treatment is usually based on the judgment of the pilots, based on visual clues and experience, but the operations director must keep the crews aware of any severe weather signatures, warnings, and the synoptic and mesoscale weather situation. Conversely, the pilots will see developing clouds before they become detectable by the radar. In general, cloud turrets producing radar echoes also already have produced significant ice and are no longer seedable. Therefore, the radar very rarely reveals the locations of treatable cells themselves, only the parent storm complexes.

The best tools for recognizing hail suppression opportunities are human eyes, a well-calibrated radar, real-time satellite imagery, and lightning detection systems, in that order. A skilled forecaster can help a great deal by alerting project personnel when storms are most likely to develop.

4.3.2.2 Treatment by Ground-Based Generators.

When ground-based generators are activated, the aerosol is produced at the surface, not within or usually near the target clouds. Convective air currents (thermals) are relied upon to carry the aerosol aloft where it can be ingested by the growing clouds. Since the thermal that generates the cloud itself is often surface-based, it is logical that seeding agent would also reach the cloud, if present when the thermal developed.

However, since the aerosol must be generated and then dispersed in the boundary layer before the cloud first develops, seeding must begin well in advance of storm development. The key to operating a ground-based ice nucleus generator network is then the successful prediction of storm development in and near the target area. Because forecasting the time of initial cloud development is difficult, most projects elect to activate ground-based generators up to several hours before cloud development is thought likely. Because forecasts are fallible, there will be times when the generator network is activated but storms do not develop.

Seeding with ground-based generators becomes less certain when mature storms move into target areas. Stronger storm updrafts may account for most of the vertical transport of the seeding agent, and seeding the main updraft is not thought to be effective (see Section 2). In addition, nocturnal storms often draw water vapor from a layer aloft, and directly from the surface. In such cases, ground-generated aerosols can simply "pool" near the surface and never reach the developing cloud turrets.

It also must be recognized that, under some circumstances, seeding by aircraft is not possible because low cloud bases and/or low visibilities render aircraft operations unsafe. In such cases, seeding with ground-based generators affords the best avenue for getting seeding agent to the desired clouds.

4.3.2.3 Treatment by Rocket or Artillery.

Rockets and artillery clearly provide the fastest means of getting seeding agent to target clouds. Delivery is "point and shoot," with the rocket or shell reaching its destination within moments. Opportunity recognition for programs employing these seeding systems is very similar to that in aircraft-based projects, except all observers are on the ground, so none have the advantage of the view from aloft. As mentioned previously, rockets and artillery have limited ranges, most no more than 12 km (7.5 mi), so many launch sites are needed. The operational infrastructure must provide for the identification of the site in best position, and establish the means by which the targeting information is to be transmitted. Some rocket-based seeding systems are computer controlled, eliminating the need for an operator to position the launcher and fire the rockets.

The difficulty with seeding by rocket is that the target clouds are not detectable by radars of the type typically employed by operational programs and, therefore, are not depicted by radar. Adjacent mature thunderstorm cells usually are detected, but their locations provide only general guidance as to where the developing turrets are located. If cloud turrets are developing 10 km (6.2 mi) from a launching site and intervening cloudiness prevents visual targeting, the correct targeting of any rockets or artillery shells fired from that site is really only by chance.

Thus, there is some uncertainty as to precisely where the ordinance should be aimed, even when suitable storms have been identified. Targeting is perhaps more certain than with ground-based generators, but still less certain than with aircraft.

4.4 COMMUNICATIONS

Communications channels among project personnel, project sponsors, and local weather officials must be open and available at all times. The most common avenues for communications are conventional telephone and two-way radio, although cellular telephone is now a viable alternative in most locations. Project personnel also should carry pagers that tell them to "phone in" for further information.

However, it is just as important that project personnel know whom to contact in any given situation, as well as how to reach them. In addition to routine operational communications (which aircraft is to launch, etc.), there are also matters of a more urgent nature, such as the need to suspend operations because of a perceived public hazard or the observation of tornadic circulations. In such cases, those who need to know must be notified immediately, without delay. It also is important that provisions be made to document all operational decisions so that, if the need ever arises, the timeliness of actions taken in the course of operations can be established.

Many programs have found that the additional severe-weather-spotting capability afforded by their radars and project personnel is well received by the local population, and appreciated by the weather offices as well. In the event that a project is based some distance from official government weather radars or observing sites, the project sponsor could notify the appropriate local civil defense entity (sheriff's department, emergency management agency, or police department) of the project, and invite them to visit and/or contact the operations center whenever they see fit. Thus, the program planners must consider how the program will be managed, and establish a reliable decision-making chain of command.

4.5 SAFETY CONSIDERATIONS

Policies and procedures that will ensure public safety and environmental integrity must be implemented. Persons of many different skills and abilities will be required to accomplish this. Many times, expertise can be drawn from beyond the project by consulting local experts. For example, knowledge of local topography and climatology is often available through the National Weather Service. In all phases, safety must be highest on the list of priorities.

4.5.1 Safety of Field Personnel
There are many potential hazards associated with hail suppression operations. These include microwave (radar) radiation; flammable and/or explosive chemicals, oxidizers, and solutions; lightning and other severe weather; and aircraft propellers. In addition, operations often continue for extended periods, tiring personnel and increasing the risk of accident.

4.5.1.1 Radar Safety.
Radars transmit microwave radiation, and so pose a significant health risk if left radiating in a fixed position at a low elevation angle for an extended period. Radars must be equipped with safety switches so that technicians can effectively disable the set when working on or near the antenna or on the transmitter or receiver. Such "lock-out, tag-out" procedures safeguard the technician and ensure that the radar is not accidentally damaged by power being applied when it should not be.

4.5.1.2 Use, Handling, and Storage of Seeding Agents.
Some chemicals used in seeding formulations, such as AgI, are not combustible in raw form and pose no fire or explosive threat. Oxidizers such as sodium perchlorate will burn vigorously at very high temperatures if ignited and may explode in confined containers. These must be stored only in approved, appropriately labeled containers in suitable locations. Pyrotechnic devices used for cloud seeding, such as burn-in-place and ejectable flares, are officially classified as Class 1.4S explosives (the same as fireworks), and must be labeled, handled, and shipped accordingly. It is illegal to ship any amount of Class 1.4S explosive on any passenger aircraft. Liquid seeding solutions are primarily acetone-based. This highly volatile chemical poses a significant fire risk, is of low viscosity (splashes easily), and produces potentially toxic vapors. Acetone solutions must be prepared and dis-

pensed only by trained personnel equipped with rubber gloves and protective eyewear. Eyewash stations should be provided to field sites where the solution is handled. When handled outdoors or in a large space (an aircraft hangar, for example), the vapors do not become concentrated to hazardous levels.

Dry ice is solid carbon dioxide, and has an unventilated surface temperature of $-78°C$ ($-108°F$). If allowed to come in contact with bare skin, freezing can result in an extremely short time. Dry ice, therefore, should be handled, sifted, and loaded into seeding equipment only by persons wearing insulated gloves. Project personnel also should be made aware that dry ice sublimates continuously at normal atmospheric pressures and temperatures and therefore displaces the oxygen needed to sustain life. Personnel should never transport dry ice within the passenger compartment of any vehicle, and all dry ice dispensing equipment onboard an aircraft must be vented to the outside only. All project personnel who will be working with seeding agents or in their proximity must be provided with MSDSs regarding each of the seeding agents so that they are fully aware of all proper and safe handling and storage requirements, as well as the potential hazards.

4.5.1.3 Severe Weather Hazards.

Hailstorms are thunderstorms. This means that each of them is capable of producing deadly lightning. Lightning is a potential hazard primarily to those on the ground, especially those working outdoors. In general, if you can hear thunder, there is some risk of cloud-to-ground lightning at your location. The nature of hail suppression operations which employ aircraft is that, on occasion, there will be circumstances when project personnel will be refueling or re-arming seeding aircraft as a thunderstorm is bearing down on an airport or as a storm is departing the vicinity. Both are high-risk situations in which the flight crew (and any others assisting on the ground) must place safety first, and delay until the storm has moved safely away. It may not take a lightning discharge to cause a tragedy, for even a static discharge at the wrong place during refueling could cause loss of aircraft and life. Since many lightning-strike victims are not killed outright, it is a good idea to provide project personnel with cardiopulmonary resuscitation (CPR) training, which could prove invaluable in many circumstances as well. The other significant weather hazard to which project personnel are often subjected is strong thunderstorm winds, which are especially dangerous to flight crews during takeoff and landing. Flight crews must resist the temptation to "beat a storm" by taking off immediately

in advance of gust fronts, and operations directors must further discourage such practices.

4.5.1.4 Aircraft Safety.

The specifics of aircraft safety during flight operations are not addressed herein, as pilots trained in seeding operations will already be cognizant of how to do so safely. However, project personnel other than flight crews will benefit from instruction on safety when around aircraft on the ground. The best safety tip is to never approach propellers (turning or still) if anyone is in the aircraft. If ground personnel are to assist flight crews at any time, the flight crews should fully brief the ground personnel on rules to observe around the aircraft.

4.5.2 Seeding Suspension Criteria

Public safety may become an issue whenever attempts are made to alter natural processes of any kind in any significant way. It is important to address scientifically based concerns and also any public perceptions (misconceptions). One means of mitigating common concerns related to excessive precipitation events has been through the use of seeding suspension criteria, wherein the presence or development of certain clouds or atmospheric conditions meeting specific criteria automatically triggers an immediate cessation of treatment.

There are many reasons why seeding might be suspended. Some hail suppression projects suspend operations if a storm becomes tornadic (see Section 3.6), or when certain weather warnings are issued by the responsible public agency. All projects should have plans to implement immediate suspension of operations whenever flash-flooding is possible, regardless of whether "official" entities have issued statements.

Seeding also may be suspended for longer periods. For example, if the target area has become excessively wet, the local sponsors may elect to stop operations, even though hail is still possible. The reasoning behind this is that seeding to suppress hail also accelerates initial precipitation development, which often results in increased rainfall. In projects sponsored by agricultural interests, hail suppression operations also may be suspended if a harvest is to be completed.

The criteria by which operations are to be suspended must be delineated in the project operations manual. This must include any graphical methods that might employ radar or precipitation observations. Special weather statements, watches, and warnings issued by the responsible governmental entities (e.g., the NWS) also can be incorporated in the suspension criteria.

4.6 PUBLIC RELATIONS, INFORMATION, AND INVOLVEMENT

If the public is not provided with factual information about the project goals, objectives, and methodologies, the result could be inaccurate speculation or deliberately misleading statements from members of the public opposed to the program. It is therefore very important that adequate information be provided as often as possible. In general, open and accessible project managers will find that most local persons will be reasonable and objective.

Many locales require hearings before a project can be undertaken or before a license or permit for the project can be granted. Concerned individuals generally will present themselves at such meetings. Their concerns should be carefully examined, and those that can be immediately and honestly addressed should be. Others may require some additional information gathering; this should be done and the concerned parties informed of the findings. Some objections, such as conflict with religious beliefs, sometimes cannot be addressed in ways that the objector finds acceptable. Concerns about adverse environmental effects from the seeding agents, excess (or insufficient) rainfall, or cost are more readily dealt with.

It can be advantageous for the program to establish a local public involvement committee, composed of responsible and open-minded citizens, if such persons can be identified and are willing to participate. Persons with extreme or inflexible views should be avoided. The local committee can serve as an avenue for concerns to be presented to the program sponsor and, hopefully, for information exchange. If the local persons can be made to understand that they have a stake in the program, and can comprehend how the program will be evaluated, they are much more likely to support it.

Informational brochures about weather modification operations, most common questions, and results can be obtained from most longer-term projects and also from the WMA. These brochures can be helpful in themselves or provide ideas for locally produced information that is specifically tailored to the project at hand.

5.0 EVALUATION OF HAIL SUPPRESSION EFFORTS

The rationale for conducting project evaluations is to find out how successful the operations are in achieving the desired results. This is more easily said than done. It is impossible to unequivocally determine the effects of seeding on the hailfall from a single convective cell. However, hail suppression projects can be justified if seeding produces a tendency toward reduced damage. Statistical evaluations are intended to identify and quantify such tendencies, but physical measurements of the cloud response also are desirable. In some projects, the sponsors have funded the deployment of a seeding aircraft equipped with microphysical instrumentation to make measurements before and after treatment. These measurements provide physical evidence of how seeding alters the cloud microstructure (Krauss 1999, Sánchez et al. 1998). Although these measurements are not made of every treated cloud, sponsors will gain confidence in the seeding system, and scientists evaluating the program will get some direct physical measurements of cloud characteristics and responses to treatment.

In this section, some of the available means for evaluation are presented. All project sponsors are strongly urged to include some evaluation procedures into their program design.

5.1 PROJECT EVALUATION APPROACHES

Most present hail suppression programs are sponsored by entities desiring the maximum immediate beneficial effect. Thus, it is rare that project sponsors are willing to leave a fraction of the potential hailstorms unseeded. Nevertheless, this is what must be done to achieve the very best scientific evaluation. However, there are other options that also may provide good evidence of program efficacy.

5.1.1 Randomized Versus Nonrandomized Programs

There are three basic approaches that may be considered for assessing the project, each with advantages and disadvantages.

1. With *randomization*, storms in the target area are seeded or not seeded on a randomized basis to develop two unbiased classes of storms for comparison.
2. With *target vs control* design, all storms in the target area are seeded, and measurements are compared with nonseeded storms in a nearby control area.
3. Projects evaluated on *historical data* compare storm measurements from within and beyond the project area both before and during the project to see if the relationship has changed because of seeding.

In the mathematical sense, the way to evaluate the efficacy of a cloud seeding program is to randomly select some storms for seeding and leave others untreated. Ideally, such randomization should be blind to those conducting the seeding, and also to those involved in the subsequent analysis. To ensure against bias in the selection of candidate clouds, the initial selector, usually a program director, makes a random choice (seed or don't seed), and relays the decision to the individual who is actually doing the seeding (the seeder), who makes a second randomized decision to either heed or ignore the first directive. Ideally, the seeder treats the cloud with either live nucleant or placebos, not knowing where each is loaded. This approach is termed a "double blind" experiment; the seeder does not know if the cloud was really treated, nor does the program director. Only after the post-analysis of the hailfall and rain, when all of the randomization decisions are combined with the known location of the seeding agent and placebo, does it become known which cases were treated and which were not. The end result of such a randomized program is more credible because the possibility of bias, conscious or unconscious, is eliminated. The drawback is that a portion of all storms is not treated, which reduces both the overall apparent effect and the benefit-to-cost ratio.

The alternative to randomized seeding is to treat every storm meeting the seeding criteria within the intended target area, while monitoring the treated storms in the target and the nontreated storms over the nearby control area. To properly evaluate such a program, it is necessary that the long-term climatologies of the storms in both areas be demonstrably similar. In other words, the proposed target and control areas must have similar precipitation histories.

5.1.2 Selection of Control Areas

The selection of the target area is made based upon the interests of the program sponsors (see Section 3.1). Once the target is determined, project evaluators must identify a nearby area that has a storm climatology as similar as possible to that of the target and that is preferably upwind to avoid inadvertent contamination by seeding agents. Thus, differences in hail damage and precipitation between the two areas will be more readily demonstrated to result from seeding operations rather than from natural variability.

5.1.2.1 Precipitation Patterns.

The historical rain and hail patterns in the target and control areas should be as similar as possible. Analyses will be based upon perceived deviations during seeding from this historical relationship, so the more these two areas are climatologically alike, the easier this will be.

5.1.2.2 Storm Frequency.

Because the target and control areas should have equal probability of hail (at least in the absence of hail suppression seeding operations), a control area should be selected that has nearly the same long-term probability of damaging hail as the target area. Usually this means that the two areas will be very close to each other and of similar character in land use, topography, and vegetation.

5.1.2.3 Contamination.

The control area must be located such that treated storms will not move from the target into the control area. Yet, to obtain optimum climatological similarity, the target and control should be as close together as possible. The control area frequently ends up being adjacent to but upwind of the target area. For evaluation purposes, any cases that may have been contaminated by seeding effects (the entrance of either seeding agent or treated storms) in the control area should be excluded from the evaluation.

5.2 EVALUATION MEASURES

There are a number of ways to approach the evaluation of hail suppression programs. The most direct way is to measure and analyze the physical properties of the precipitation intended to be affected by the operations, e.g., the hail and rain reaching the surface. Another means is to study secondary data sets such as crop-hail insurance loss claims. The economic side of operations can be evaluated by modeling the impacts of observed or predicted changes in hail losses. An example of this would be changes in crop-hail insurance claims. Yet another indication of program success might result from careful examination of crop yield data. If the program is successful, the average production should increase. Finally, some indications of the efficacy of operations also can be gained through the use of numerical (computer) cloud models that simulate the effects of seeding. Historically, simulations have generally been reserved for programs having significant research components. However, they would likely prove beneficial to any program, especially if both seeded and nonseeded storms had a model component.

5.2.1 Evaluations using Direct Evidence

Direct measurement of hail and rain will afford the best basis for project evaluation. While obtaining these measurements in sufficient quantity and quality is difficult and costly, serious consideration should be given to this aspect.

5.2.1.1 Hail Pad Networks.

The best method currently available to quantify surface hailfall is to use a hail pad network. A hail pad is a square of material placed horizontally in an exposed area (Schleusener and Jennings 1960). The pad is generally composed of a firm foam of some type that will deform permanently when struck by hail (Figure 5-1). To protect from degradation of the foam by UV sunlight, the pad should either be covered with a thin layer of aluminum foil or painted with a thin layer of UV-blocking latex paint. Hail pads are routinely checked for evidence of hail impacts. When such impacts are recorded, the foam square is labeled and removed for analysis and replaced with a new square. Because hail is a highly variable phenomenon with respect to both space and time, the best results for evaluation are obtained when hail pads are deployed in regular grid-spacings over large areas. For example, a hail pad network consisting of 150 pads deployed in a 10 by 15 array, with each pad 1 km (0.6 mi) from its neighbors, would have a reasonable chance of detecting modest-to-large hail streaks through the area covered by the network. Hail pad networks are extremely labor-intensive. They require regular inspection after each storm and also during periods of nonconvective weather to ensure all pads are properly positioned and undamaged. Pads receiving hail damage must be numbered, removed, replaced, and then analyzed for the number and sizes of impact marks. Prior to 1985, hail pads had to be laboriously and subjectively analyzed by hand; but there now exist automatic means for objective analysis that employ various digital imaging techniques. Generally, a hail pad network requires a full-time crew during the hail season, and a like amount of time for post-analysis. In spite of this, hail pad networks are possibly one of the least expensive sources of direct evidence of hail size and energy. In Europe, hail pad networks have been employed successfully in Spain (Fraile et al. 1991, 1992), France (Dessens 1998), Italy, and Greece. In some cases, the networks have been used to evaluate non-randomized programs (Dessens et al. 1998). The data provided are also useful in hail research.

5.2.1.2 Rainfall Data.

The best measurements of rainfall are obtained through surface rain gauges. The quality of each individual measurement depends upon the size of the orifice of the gauge, gauge shielding (if any), gauge siting, and whether the precipitation is time-resolved. Standard recording rain gauges with 8-in. orifice are

FIGURE 5-1. A foil-covered hail pad after exposure to small hail. Exposed pad area is about 1,000 cm² (1.1 ft²). (Courtesy P. Simeonov, National Institute of Meteorology and Hydrology, Sofia, Bulgaria.)

well suited to this task but are expensive and require attention on the same order as hail pads. Few quality reporting sites are generally found in the desired areas. Thus, a project interested in using direct rainfall measurements for evaluations must significantly augment existing reporting sites. This must be done in both the target and control areas. This technique is only applicable to programs that adopt randomization, since the target vs control approach requires a historical, non-seeded data set.

5.2.1.3 Radar Data.

Today's radar hardware and software provide a means by which larger-scale cloud behaviors can be documented with relatively little effort. Data recording systems such as the TITAN record reflectivity in a volume-scan mode, estimate storm-integrated precipitation, and perform a wide variety of real-time tasks helpful in directing operations, such as user-selectable vertical cross sections, vertically integrated liquid (VIL) cloud water, and projected storm tracks. A radar positioned to provide comparable areal and range coverage for both target and control can be very helpful in evaluations by documenting radar-estimated rain volumes for both seeded and unseeded storms.

Most portions of the United States are now adequately covered by the NWS WSR-88D NEXRAD Doppler radar network. Data from these radars are archived and available through the National Climatic Data Center. This database may be helpful for the aspects of project design that depend upon the storm climatology. However, it is currently of little use for decision-making in real-time because, at most projects' operations centers, it is available only via the Internet in base reflectivity scans that are updated once every half-hour or so. The future holds considerable hope for project analysis and evaluation through the NWS radar network when it has dual-polarization capability added, which is scheduled to occur sometime in the next decade. This capability has been shown to provide the ability to differentiate between ice-phase and liquid water-phase hydrometeors (Reinking et al. 1998). Thus, dual-polarization capability should become an excellent tool for the future assessment of hail suppression operations.

Although radar is very helpful for documenting storm behavior, its use for quantitative evaluation of hail suppression efforts is nontrivial. Storms are viewed by radar differently, depending upon distance from the radar (range), viewing aspect, and the possible presence of other intervening clouds and precipitation. Thus, quantifying differences between two storms, one seeded and one unseeded, that are at different ranges and viewed from different aspects becomes very complex.

5.2.2. Evaluation Through Secondary Evidence

Secondary evidence, that is, data that reflect program efficacy but that are not direct measures of it, may be very useful. Insurance loss-cost ratios and crop yield statistics are two examples of secondary evidence. Both can provide evidence of program efficacy.

5.2.2.1. Crop-Hail Insurance Data.

Many hail suppression programs are sponsored for the express purpose of reducing crop-hail damage. Thus, examination of crop-hail data for evidence of seeding effects (e.g., Smith et al. 1997) seems obligatory. Data for much of the United States are available from the National Crop Insurance Services. Considerable care must be taken to ensure that the policy forms used were the same in both target and control areas through the period of interest, and that representative liability was sold in each. This is important because any uninsured crop could be a total loss due to hail, and it would never show up in the crop-loss data. The vulnerability of crops to hail varies not only with hail size and wind, but also with the crop type, its health, and its stage of development. In some cases, hailfall occurs where total crop loss has previously occurred, in which case the latter hailfall would go undetected by crop-hail insurance data. Thus, the use of crop-hail data must be undertaken with considerable care and results of such evaluations presented with the appropriate caveats.

5.2.2.2. Insurance Property Loss Data.

Hail damage to property has become a major problem, with the first $1.1 billion hailstorm now on record that occurred in Dallas–Ft. Worth, Texas, in 1995 (Institute for Business and Home Safety (IBHS) 1998). Storms causing $100 million or more in damages are not uncommon. For example, storms causing in excess of $300 million occurred during the 1990s in Orlando, Wichita, Dallas, and Oklahoma City (Changnon 1997). However, property and casualty losses are recorded and reported differently among companies, and, unlike crop-hail insurance, quality statistics are not available for the industry. Statistics that are available from individual companies are seldom reported by area and date, and many include all storm damage, not just hail. Thus, getting meaningful data is currently very problematic. The insurance industry is becoming increasingly aware of the need for better reporting, but most companies consider their data proprietary, fearing the competition might gain an advantage if data are divulged. The IBHS has considered serving as an anonymous clearinghouse for hail-loss information, but this would require further standardization of reporting, which would be an expensive conversion for many companies. Therefore, property hail-loss statistics are not currently available. Even though hail-loss reduction in a large urban area would be extremely beneficial, meaningful statistics on total present property losses are not available.

5.2.2.3. Crop Yield Data.

The National Agricultural Statistics Service (NASS) publishes crop yield statistics for most of the nation. These numbers, if carefully examined, can be used to approximate project benefits. In general, their use requires a long-term project of at least ten growing seasons. Target and control areas are required that have well-established, long-term historical cropping practices and similar yields in nonseeded periods. Other factors that might explain observed differences must be carefully screened. Among these are changes in farming practices in either the target or control area. These possible changes include irrigation, use of agrichemicals, and seed varieties. Observed changes in crop yields can be used as an indicator of program efficacy but not as a precise measure of it. A positive change in yield should be encouraging, and a negative change would be cause for concern. The advantage of using crop-yield statistics is that they are readily available, relatively inexpensive, and regularly compiled. Obtaining historical data also is generally not problematic.

5.3 DISSEMINATION OF RESULTS

The statistical evaluation of any program can be strengthened by physical measurements of the clouds and precipitation, so that a physical basis (explanation) of the observed effects can be established. Whether or not physical measurements are made, the natural variability of hail is such that a sound evaluation will require the project to run for a number of years, preferably ten or more. Project sponsors must be aware of this from the beginning. An understanding should be established as to how "preliminary" evaluations might be used to get a sense of program efficacy. Comparisons of secondary data in the target and control areas of the types listed in Section 5.2 could be a start. Such preliminary results often can be helpful from the standpoint of encouraging the sponsor; however, they are very seldom statistically significant.

In the early years of a new program, after the conclusion of each season it is a good idea to release statements to the media concerning perceived program efficacy. Appropriate caveats must be included.

As data accrue over several years, the results gain certainty, and confidence increases accordingly. Publication of these findings should follow, preferably in refereed scientific journals. If there is interest at any nearby universities, researchers there should become involved, particularly if objectivity can be maintained.

6.0 GLOSSARY OF TERMS AND ACRONYMS

Definitions are from the Glossary of Meteorology (AMS 1959, 2000), where applicable. Related glossary entries are shown in *italic type*.

accumulation zone—hypothetical, relatively small regions within mature updrafts of hailstorms where supercooled liquid water drops have accumulated, creating a region well-suited to the rapid development of large hail. Although such zones have been encountered by research aircraft on rare occasions, their true role, if any, in hail development is uncertain.

AgI—see *silver iodide*.

AMS—American Meteorological Society, 45 Beacon Street, Boston, MA 02108-3693.

ASCE—American Society of Civil Engineers, 1801 Alexander Bell Drive, Reston, VA 20191-4400.

attenuation—in physics, any process in which the flux density (or power, amplitude, intensity, illuminance, etc.) of a "parallel beam" of energy decreases with increasing distance from the energy source, for example, the reduction of intensity of the electromagnetic wave (radar signal) along its path from and back to the radar. Attenuation thus lessens the ability of a radar to sense all clouds and precipitation, such that the depicted information is inaccurate or incomplete.

AZ—see *accumulation zone*. Also used to indicate azimuth, especially in radar manuals.

broadcast seeding—the release of seeding agent, either from aircraft or from the ground, in conditions thought favorable for the development of treatable convective storms, but either before such storms have developed or at some distance from the storms (i.e., not within the storm or in its immediate proximity). Compare *direct targeting*. Broadcast seeding is also a routine practice to increase precipitation within winter clouds in mountainous areas.

buffer zone—the area between the target area and control area that buffers the control from the possible effects of seeding within the operational area.

burn-in-place flare—a pyrotechnic device burned in a fixed position, such as the trailing edge of an aircraft wing. Compare *ejectable flare*.

CCN—cloud condensation nuclei. The tiny particles, either liquid or solid, upon which condensation of water vapor first begins in the atmosphere; necessary for the formation of cloud droplets.

CD-ROM—compact disk, read-only memory. Much of the project data is archived to CD-ROM because most modern computers can read them.

cell—a convective element (cloud) that, in its life cycle, develops, matures, and dissipates, usually in about 30 min.

certified manager—certification of weather modification project managerial experience and skills granted by the *WMA*.

certified operator—certification of weather modification project operational experience and skills granted by the *WMA*.

cloud droplets—a particle of liquid water from a few microns to tens of microns in diameter, formed by condensation of atmospheric water vapor, and suspended in the atmosphere with other droplets to form a cloud. These liquid water droplets are too small to precipitate.

cloud model—physical description of cloud processes programmed into a computer to simulate cloud development and evolution. Useful in understanding the relative importance of the many factors that influence cloud development, and the only way in which *exactly the same cloud* can be both seeded and unseeded.

CO₂—see *dry ice*.

coalescence—in cloud physics, the merging of two water drops into a single larger drop. This occurs through the collision of two drops, which then unite.

conceptual model—a theoretical model of hail development, based upon current knowledge and scientific concepts. See also *cloud model*.

contamination—the inadvertent distribution of seeding agent into areas that, according to project design, were not to have been seeded. This becomes a problem only when such areas contain storms, which then can no longer be considered unseeded.

control area—an area where cloud seeding operations do not take place, preferably similar in character and near to the *target area*. The behavior of storms over the control area is compared to treated storms over the target area to assess differences and thus measure project effectiveness. See also *target area*, *operational area*, and *buffer zone*.

crossover design—a project that employs areas that alternate between target and control. Crossover re-

duces the possibility of geographically induced bias in the evaluation.

direct targeting—the placement of seeding agents directly into the target cloud mass, either by release during penetration by aircraft, rocket, or artillery, or from aircraft flying directly below cloud base in updraft. Compare *broadcast seeding*.

droplet spectrum—the numbers and sizes of the droplets within the cloud volume of interest.

dry growth—growth of a hailstone or hail embryo in supercooled cloud having a relatively low liquid water content, such that supercooled cloud droplets freeze upon impact, resulting in accretion that is opaque and porous.

dry ice—frozen carbon dioxide (CO₂). Dry ice pellets have an unventilated surface temperature of −78°C (−108°F), and have been used considerably for glaciogenic cloud seeding. Ventilated surface temperatures, such as might be realized during free-fall within a cloud, are near −105°C (−157°F).

dynamic seeding—the treatment of clouds with the intent of using the latent heat produced by additional freezing and perhaps in some cases by condensation or deposition to invigorate cloud development.

ejectable flares—pyrotechnic devices that are ignited and released (ejected) from aircraft. Compare *burn-in-place flare*.

embryo—see *hail embryo*.

FAA—Federal Aviation Administration. The government entity that regulates aircraft operations, safety, and use of airways in the United States. Analogous entities also exist in most other nations.

flanking line—developing convective cells on the flank (side) of a mature thunderstorm.

glaciogenic—causing the formation of ice.

glaciogenic seeding—treatment of clouds with materials intended to increase and/or initiate the formation of ice crystals.

GOES—Geostationary Operational Environmental Satellite. These are the latest NOAA weather satellites, currently operational over the continental United States.

GPS—Global Positioning System. A global, multiple-satellite-based navigation positioning system that provides consistently accurate positions.

graupel—white, opaque, approximately round (sometimes conical) ice particles having a snow-like structure, and about 2 to 5 mm (0.08 to 0.20 in.) in diameter. Also known as snow pellets, they form in convective clouds when supercooled water droplets freeze to an ice particle upon impact.

Grossversuch IV—a Swiss hail suppression research program conducted from 1977 to 1981, which

49

attempted to replicate the reported successes of the former Soviet Union's hail suppression program. See also *NHRE*.

hail—precipitation in the form of balls or irregular lumps of ice, always produced by convective clouds, nearly always by cumulonimbus. By convention, hail has a diameter of 5 mm (0.20 in.) or more.

hail cannon—a vertically pointing device that uses explosions to attempt to prevent hail development. Though no scientific basis for the functionality of hail cannons has been published in refereed literature, their use continues in a few places, usually in high-value, limited area agricultural endeavors, such as orchards.

hail embryo—small ice particles, often graupel or frozen raindrops that, when resident in an updraft of sufficient supercooled liquid water content, grow into hailstones.

hail pad—a device made of a relatively soft but nonresilient material, such as Styrofoam, that deforms when impacted by hailstones. The size and depth of the deformations are used to estimate the kinetic energy and size of the hailstones. See also *UV*.

hailstone—an individual unit of hail.

hail streak—a deposit of hail on the ground. Sometimes used interchangeably with *hail swath,* a hail streak also may be a region of greater damage within a hail swath, presumably due to greater sizes or numbers of hailstones.

hail swath—the surface path of hail at the surface. See also *hail streak.*

hydrometeor—any product of condensation or deposition, or condensation and freezing, in the atmosphere. This includes cloud water or ice of any size, either suspended in the air or precipitating.

hygroscopic—pertaining to a marked ability to accelerate the condensation of water vapor; having the property of attracting water, or having the effect of encouraging the formation of larger droplets.

hygroscopic seeding—treatment of clouds with hygroscopic materials that encourage the formation of larger droplets, changing the cloud droplet spectrum to enhance development of precipitation through coalescence.

IBHS—Institute for Business and Home Safety (formerly the Insurance Institute for Property Loss Reduction), 1408 North Westshore Boulevard, Suite 208, Tampa, FL 33607.

ice nucleus (IN)—any particle that serves as a nucleus for the formation of ice crystals in the atmosphere.

IFF—Identification, Friend or Foe. An older aircraft tracking system that interrogates aircraft L-band transponders. The IFF systems have largely been replaced with GPS-based systems, which are considerably more precise and also more reliable.

IFR—Instrument Flight Rules. The FAA regulations pertaining to flight at altitudes of 18,000 ft (5.5 km) above mean sea level or higher over U.S. airspace, or in any meteorological conditions necessitating the use of aircraft instrumentation for safe navigation.

IN—see *ice nucleus.*

in situ measurement—measurements made in place, as within the cloud of interest. Compare *remote sensing.*

KCl—see *potassium chloride.*

latent heat—the heat released or absorbed per unit mass by a system in a reversible, isobaric-isothermal change of phase. More simply, the heat released when water vapor condenses (latent heat of condensation), or when liquid water drops freeze (latent heat of fusion). In the case of water droplets freezing upon contact with hail, the latent heat elevates the surface temperature of the growing hailstone.

loss-cost ratio—in insurance, the ratio of loss to the liability, multiplied by 100. For example, a $20 loss on a policy having an insured liability of $40 would be a loss-cost of 50%.

METEOSAT—the current European weather satellites. See also *GOES,* METEOSAT's counterpart in the United States.

NaCl—see *sodium chloride.*

NASS—National Agricultural Statistics Service, 5285 Port Royal Road, Springfield, VA 22161.

NCAR—National Center for Atmospheric Research, P.O. Box 3000, Boulder, CO, 80307-3000.

NEXRAD—see *WSR-88D.*

NHRE—National Hail Research Experiment. Conducted in the early 1970s in northeast Colorado, it attempted to replicate the reported successes of the former Soviet Union's hail suppression program. See also *Grossversuch IV.*

NOAA—National Oceanic and Atmospheric Administration, U.S. Department of Commerce. The parent organization of the U.S. National Weather Service, and the federal agency to which all U.S. weather modification activities must be reported.

nowcasting—very short-term forecasting, from the present to about 30 min.

NWS—National Weather Service, the United States public weather forecasting agency. The NWS is a division of NOAA.

operational area—the area over or within which seeding operations are actually conducted, including the target area and usually adjacent areas beyond the boundaries of the target area, where seeding is conducted with the intent of affecting clouds over the target.

overseeding—a condition resulting from the application of too much glaciogenic seeding agent in which too many small ice crystals may form, none of which is large enough to precipitate or aggregate.

placebo—treatment with an inert substance, without the knowledge of those applying the treatment. In a randomized cloud seeding program, clouds are treated with real seeding agents or a placebo, which might be only an audible event such as a recorded "bang" that sounds like a flare firing or a flare containing sand.

potassium chloride—KCl, a simple salt often used as a primary ingredient in hygroscopic cloud seeding pyrotechnics.

raindrop—a drop of water of diameter greater than 0.5 mm (0.20 in.) falling through the atmosphere. In careful usage, falling drops with diameters lying in the interval 0.2 to 0.5 mm (0.08 to 0.20 in.) are called "drizzle drops" rather than raindrops, although this is frequently overlooked.

rawinsonde (or radiosonde)—an instrument package that senses and transmits weather information such as pressure, temperature, and humidity. Rawinsondes are carried aloft by weather balloons twice daily from numerous sites all over the world; also can be employed by projects to bolster local forecasting efforts.

remote sensing—the remote measurement of properties of interest, as with radar and satellite. Compare *in situ measurement*.

response time—the time that elapses from identification of a seeding opportunity until the release of seeding agent actually begins.

seeding agents—agents dispensed by any means in or near a cloud volume that are intended to modify (seed) the cloud characteristics.

silver iodide—AgI, a common glaciogenic seeding agent.

sodium chloride—NaCl, the chemical composition of common table salt. Because of its hygroscopic properties, historically it has been used occasionally for hygroscopic seeding. Hygroscopic seeding agents have more recently employed *potassium chloride* (KCl).

supercell—thunderstorms characterized by an intense, quasi–steady-state mature updraft. Such storms account for a large fraction of all tornadoes, and much of the large hail.

supercooled water—water, still in liquid state, at temperatures less than 0°C (32°F). Under ideal conditions in the free atmosphere, water may exist in a supercooled state to temperatures as cold as −40°C (−40°F).

target area—the area for which cloud seeding operations are targeted, usually near a *control area* that is similar in character and climatology. The behavior of treated storms over the target area is compared to untreated storms over the control area, to assess differences and thus measure project effectiveness. See also, *control area*, *operational area*, and *buffer zone*.

terminal velocity—the particular falling speed, for any given object moving through a fluid of specified physical properties, at which the drag forces and buoyant forces exerted by the fluid on the object just equal the gravitational force acting on the object. For *hydrometeors*, it is the greatest fall speed relative to the surrounding air that a hydrometeor will attain, as determined by the mass of the particle and frictional drag of the air through which it is falling.

thermal—a relatively small-scale, rising current of air produced when the atmosphere is heated enough locally by the earth's surface to produce absolute instability in the lowest layers.

TITAN—Thunderstorm Identification, Tracking, Analysis, and Nowcasting. Software for the display and analysis of weather radar data, widely used in operational convective cloud seeding programs.

UV—ultraviolet electromagnetic radiation of shorter wavelength than visible radiation, but longer than x-rays. Ultraviolet radiation (light) is a component of normal solar radiation. Foam *hail pads* will degrade with prolonged exposure to UV radiation, and so are either covered with foil or painted.

VIL—vertically integrated liquid. A radar estimate of the cloud liquid water, from the lowest angle sampled through cloud top. Used as an indicator of the presence of hail.

wet growth—growth of a *hailstone* or *hail embryo* in supercooled cloud having a liquid water content such that the latent heat of freezing warms the surface to near 0°C (+32°F), so that all supercooled cloud droplets do not immediately freeze upon impact. This results in ice growth that is often transparent and nearly solid.

wing-tip generator—ice nucleus generators mounted at the tips of aircraft wings, or sometimes below the wings, also usually near the ends.

WMA—Weather Modification Association, P.O. Box 26926, Fresno, CA 93729-6926.

WMO—World Meteorological Organization, 7 bis, Avenue de la Paix, CH 1211 Geneva 2, Switzerland.

WSR-88D (NEXRAD)—the 1988 vintage Doppler weather radar network deployed in the United States by the National Weather Service during the 1990s.

7.0 REFERENCES

Admirat, P., 1972: Diffusion théorique et expérimentale de noyaux d'argent émis par des réseaux de generatours. *J. Rech. Atmos.*, **6**, 13–27.

AMS, 1959: *Glossary of meteorology*, R.E. Huschke, ed. American Meteorological Society, Boston. 638 p. (Revised 1970.)

AMS, 1998: Policy statement: planned and inadvertent weather modification. *Bull. Amer. Meteor. Soc.*, **79**, 2771–2772.

AMS, 2000: *Glossary of meteorology*, 2nd ed., Todd S. Glickman, ed. American Meteorological Society, Boston, 855 p.

Appleman, H., 1958: An investigation into the formation of hail. *Nubila*, **1**, 28–37.

Atlas, D., 1965: Activities in radar meteorology, cloud physics, and weather modification in the Soviet Union. *Bull. Amer. Meteor. Soc.*, **46**, 696–706.

Battan, L.J., 1965: A view of cloud physics and weather modification in the Soviet Union. *Bull. Amer. Meteor. Soc.*, **46**, 309–316.

Boe, B.A., 1992: Hail suppression in North Dakota. Preprints, AMS Symposium on Planned and Inadvertent Weather Modification, Atlanta, GA, 58–62.

Boe, B.A., 1994: The North Dakota tracer experiment: tracer applications in a cooperative thunderstorm research program. *J. Weather Modif.*, **26**, 102–112.

Boe, B.A., D.L. Langerud, and P. Moen, 1998: North Dakota cloud modification project operations manual. North Dakota Atmospheric Resource Board, Bismarck, 61 p.

Boe, B.A., J.L. Stith, P.L. Smith, J.H. Hirsch, J.H. Helsdon, Jr., A.G. Detwiler, H.D. Orville, B.E. Martner, R.F. Reinking, R.J. Meitín, and R.A. Brown, 1992: The North Dakota thunderstorm project: a cooperative study of High Plains thunderstorms. *Bull. Amer. Meteor. Soc.*, **73**, 145–160.

Bruintjes, R.T., D.W. Breed, B.G. Foote, M.J. Dixon, B.G. Brown, V. Salazar, and H.R. Rodriguez, 1999: Program for the augmentation of rainfall in Coahuilla (PARC): overview and design. 7th WMO Scientific Conf. on Weather Modification, Chiang Mai, Thailand, February 17–22, 1999. WMO Technical Document No. 936, WMP Report No. 31, 53–56.

Bruintjes, R.T., T.L. Clark, and W.D. Hall, 1995: The dispersion of tracer plumes in mountainous regions in central Arizona: comparisons between observations and modeling results. *J. Appl. Meteor.*, **34**, 971–988.

Carte, A.E., R.H. Douglas, R.C. Srivastava, and G.N. Williams, 1963: Alberta hail studies. Rep. MW-36, McGill University, Montreal.

Castro, A., J. L. Marcos, J. Dessens, J.L. Sánchez, and R. Fraile, 1998: Concentration of ice nuclei in continental and maritime air masses in León (Spain). *Atmos. Res.*, **47–48**, 155–168.

Changnon, S.A., 1977: The climatology of hail in North America, in *A review of hail science and hail suppression*, G.B. Foote and C.A. Knight, eds. *Meteorological Monographs*, **16**(38), 107–126.

Changnon, S.A., 1997: Trends in hail in the United States. Proceedings, workshop on the social and economic impacts of weather. National Center for Atmospheric Research, Boulder, CO, 19–34.

Changnon, S.A., and J.L. Ivens, 1981: History repeated: the forgotten hail cannons of Europe. *Bull. Amer. Meteor. Soc.*, **62**, 368–375.

Cooper, W.A., and J. Marwitz, 1980: Winter storms over the San Juan Mountains, part III, seeding potential. *J. Appl. Meteor.*, **19**, 942–949.

Cooper, W.A., R.T. Bruintjes, and G.K. Mather, 1997: Calculations pertaining to hygroscopic seeding with flares. *J. Appl. Meteor.*, **36**, 1449–1469.

DeMott, P.J., 1991: Comments on "The persistence of seeding effects in a winter orographic cloud seeded with silver iodide burned in acetone." *J. Appl. Meteor.*, **30**, 1376–1380.

DeMott, P.J., A.B. Super, G. Langer, D.C. Rogers, and J.T. McPartland, 1995: Comparative characterizations of the ice nucleus ability of AgI aerosols by three methods. *J. Weather Modif.*, **27**, 1–16.

Dennis, A.S., 1977: Hail suppression concepts and seeding methods, in *Hail: a review of hail science and hail suppression,* G.B. Foote and C.A. Knight, eds. *Meteorological Monographs*, **16**(38), 181–191.

Dennis, A.S., 1980: *Weather modification by cloud seeding.* Academic Press, New York, 267 p.

Dennis, A.S., and A. Koscielski, 1972: Height and temperature of first echoes in unseeded and seeded convective clouds in South Dakota. *J. Appl. Meteor.*, **11**, 994–1000.

Dessens, J., 1986: Hail in southwestern France, part II: results of a 30 year hail suppression project with silver iodide seeding from the ground. *J. Appl. Meteor.*, **25**, 45–58.

Dessens, J., 1998: A physical evaluation of a hail suppression project with silver iodide ground generators in southwestern France. *J. Appl. Meteor.*, **37**, 1588–1599.

Dessens J., J.L. Sánchez, and R. Fraile, 1998: An analysis of the geographical disposal of silver iodide ground generators for hail prevention. Proceedings of 13th AMS Conf. on Planned and Inadvertent Weather Modification, Everett, Washington, August 17–21, 1998, 585–588.

Dixon, M., and G. Weiner, 1993: TITAN, thunderstorm identification, tracking, analysis, and nowcasting: a radar-based methodology. *J. Atmos. Ocean. Tech.*, **10**, 785–797.

Dye, J.E., A.J. Heymsfield, R.R. Paluch, and D.W. Breed, 1976: Final report, national hail research experiment randomized seeding experiment, 1972–1974, vol. II. National Center for Atmospheric Research, Boulder, CO, 530 p.

Farley, R.D., and H.D. Orville, 1999: Whence large hail? 7th WMO Scientific Conf. on Weather Modification, Chiang Mai, Thailand, February 17–22, 1999. WMO Technical Document No. 936, WMP Report No. 31, 507–510.

Farley, R.D., H. Chen, H.D. Orville, and M.R. Hjelmfelt, 1996: The numerical simulation of the effects of cloud seeding on hailstorms. Preprints, 13th AMS Conf. on Planned and Inadvertent Weather Modification, Atlanta, GA, 23–30.

Federer, B., 1977: Methods and results of hail suppression in Europe and the U.S.S.R. In *Hail: a review of hail science and hail suppression,* G.B. Foote and C.A. Knight, eds. *Meteorological Monographs*, **16**(38), 215–224.

Federer, B., A. Waldvogel, W. Schmid, H.H. Schiesser, F. Hampel, M. Schweingruber, W. Stahel, J. Batter, J.F. Meziex, N. Doras, G. D'Aubingny, G. Der-Megreditchian, and D. Vento, 1986: Main results of Grossversuch IV. *J. Clim. and Appl. Meteor.*, **26**, 917–957.

Finnegan, W.G., 1998: Rates and mechanisms of heterogeneous ice nucleation on silver iodide and silver chloroiodide particulate substrates. *J. Colloid Interface Sci.*, **202**, 518–526.

Finnegan, W.G., A.B. Long, and R.L. Pitter, 1994: Specific applications of ice nucleus aerosols in weather modification. 6th WMO Conf. on Weather Modification, Pæstum, Italy. WMO Technical Document No. 596, WMP Report No. 22, 247–250.

Foote, G.B., and C.A. Knight (eds.), 1977: *Hail: a review of hail science and hail suppression. Meteorological Monographs*, **16**(38), 277.

Foote, G.B., and C.A. Knight, 1979: Results of a randomized hail suppression experiment in northeast Colorado, part I: design and conduct of the experiment. *J. Appl. Meteor.*, **18**, 1526–1537.

Fraile R., A. Castro, and J.L. Sánchez, 1992: Analysis of hailstone size distributions from a hail pad network. *Atmos. Res.*, **28**, 331–326.

Fraile R., J.L. Sánchez, J.L. de la Madrid, and A. Castro, 1991: A network of hail pads in Spain. *J. Wea. Modif.*, **23**, 56–62.

Garvey, D.M., 1975: Testing of cloud seeding materials at the Cloud Simulation and Aerosol Laboratory, 1971–1973. *J. Appl. Meteor.*, **14**, 883–890.

Grandia, K.L, D.D. Davison, and J.H. Renick, 1979: On the dispersion of silver iodide in Alberta hailstorms. Preprints, 7th AMS Conf. on Planned and Inadvertent Weather Modification, Banff, Alberta, Canada, 56–57.

Griffith, D.A., G.W. Wilkerson, and D.A. Risch, 1990: Airborne observation of a summertime ground-based tracer gas release. *J. Weather Modif.*, **22**, 43–48.

Heimbach, J.A., Jr., and N.C. Stone, 1984: Ascent of surface-released silver iodide into summer convection, Alberta 1975. *J. Weather Modif.*, **16**, 19–26.

Hitschfeld, W., 1971: Hail research at McGill, 1956–1971. Stormy Weather Group, Sci. Rep. MW-68. McGill University, Montreal.

Hohl, R., H. Scheisser, and I. Kriepper, 2002: The use of weather radars to estimate hail damage to automobiles: an exploratory study in Switzerland. *Atmos. Res.*, **61**, 215–238.

Holroyd, E.W. III, A.B. Super, and B.A. Silverman, 1978: The practicability of dry ice for on-top seeding of convective clouds. *J. Appl. Meteor.*, **17**, 49–63.

IBHS, 1998: *The insured cost of natural disasters: a report on the IBHS paid loss data base.* Institute for Business and Home Safety, Tampa, FL, 24 p.

Kahan, A.M., D. Rottner, R. Sena, and C.G. Keyes, Jr., 1995: *Guidelines for cloud seeding to augment precipitation.* ASCE Manuals and Reports on Engineering Practice No. 81. ASCE, New York, 145 p.

Klein, A.D., 1978: *Environmental impacts of artificial ice nucleating agents.* Dowden, Hutchinson & Ross, Inc., Stroudsburg, PA, 218 p.

Knollenberg, R.R., 1970: The optical array: an alternative to scattering and extinction for airborne particle size determination. *J. Appl. Meteor.*, **9**, 86–103.

Knollenberg, R.R., 1976: Three new instruments for cloud physics measurements: the 2-D spectrometer, the forward-scattering spectrometer probe, and the active scattering aerosol spectrometer. Preprints, Intl. AMS Conf. on Cloud Physics, Boulder, CO, 554–561.

Kopp, A.J., and H.D. Orville, 1994: The use of a two-dimensional, time-dependent cloud model to predict convective and stratiform clouds and precipitation. *J. Wea. Forecast.*, **9**, 62–77.

Korolev, A.V., G.A. Isaac, and J. Hallett, 1999: Ice particle habits in Arctic clouds. *Geophys. Res. Letters*, **26**(9), 1299–1302.

Krauss, E.W., and J. Renick, 1997: The new Alberta hail suppression project. *J. Weather Modif.*, **29**, 100–105.

Krauss, T.W., 1999: Mendoza hail suppression project: final report executive summary. Ministerio de Economy, Gobierno de Mendoza, Argentina. Weather Modification, Inc., Fargo, ND, 130 p.

Lawson, R.P., B.A. Baker, and C.G. Schmitt, T.L. Jensen, 2001: Overview of microphysical properties of Arctic stratus clouds observed in May and July during FIRE.ACE. *J. Geophys. Res.* **106,** 1001–1042.

Lee, A.C.L., 1986: An operational system for the remote location of lightning flashes using VLF arrival time difference technique. *J. Atmos. Ocean. Tech.*, **6,** 43–49.

Linkletter, G.O., and J.A. Warburton, 1977: An assessment of NHRE hail suppression seeding technology based on silver analysis. *J. Appl. Meteor.*, **16,** 1332–1348.

List, R., 1963: On the effect of explosive waves on hailstone models. *J. Appl. Meteor.*, **2,** 494–497.

Lominadze, V.P., I.T. Bartishvili, and S.L. Gudushauri, 1973: On the results of practical protection of valuable agricultural crops from hail by the THRI (Zaknigmi) method. Proceedings, WMO/IAMAP Scientific Conf. on Weather Modification, Tashkent. WMO Technical Document No. 399, 225–230.

Ludlam, F.H., 1958: The hail problem. *Nubila*, **1,** 12.

MacGorman, D.R., and D.W. Burgess, 1994: Positive cloud-to-ground lightning in tornadic storms and hailstorms. *Mon. Weather Rev.*, **122,** 1671–1697.

MacGorman, D.R., and W.D. Rust, 1998: *The electrical nature of storms*. Oxford University Press, New York, 422 p.

Mather, G.K., 1991: Coalescence enhancement in large multicell storms caused by the emissions of a Kraft paper mill. *J. Appl. Meteor.*, **30,** 1134–1146.

Mather, G.K., and D. Terblanche, 1994: Initial results from cloud seeding experiments using hygroscopic flares. 6th WMO Conf. on Weather Modification, Pæstum, Italy. WMO Technical Document No. 596, WMP Report No. 22, 687–690.

Mather, G.K., M.J. Dixon, and J. M. deJager, 1996: Assessing the potential for rain augmentation: the Nelspruit randomized convective cloud seeding experiment. *J. Appl. Meteor.*, **35,** 1465–1482.

Mather, G.K., D.E. Terblanche, F.E. Steffens, and L. Fletcher, 1997: Results of the South African cloud-seeding experiments using hygroscopic flares. *J. Appl. Meteor.*, **36,** 1433–1447.

McPartland, J.T., and A.B. Super, 1978: Diffusion of ground-generated silver iodide to cumulus formation levels. *J. Weather Modif.*, **10,** 71–76.

Mezeix, J.F., M. Segarra, J.L. Tondut, and B. Vaissiers, 1974: Etude preliminaire d'un canon detanant pour la prevention de la grele. Report of Groupment Interdepartmental d'Etudes des Fleaux Atmospheriques, Valance, France.

Morgan, G.M., 1973: A general description of the hail problem in the Po Valley of northern Italy. *J. Appl. Meteor.*, **12,** 338–353.

Morgan, G.M., 1982a: Observation and measurement of hailfall. Chapter 8 in *Thunderstorms: a social, scientific, and technological documentary, vol. 3, instruments and techniques for thunderstorm observation and analysis*, E. Kessler (ed.). NOAA, Environmental Research Laboratories, 149–180.

Morgan, G.M., 1982b: Precipitation at the ground. Chapter 4 in *Hailstorms of the Central High Plains, vol. I: the national hail research experiment*. C.A. Knight and P. Squires, eds. Colorado Associated University Press, Boulder, CO, 391 p.

Morgan, G.M., Jr., D. Vento, J.F. Mezeix, P. Admirat, B. Federer, A. Waldvogel, L. Wojtiv, and E. Wirth, 1980: A comparison of hailfall intensity and spatial variability in the United States and Europe based on data from instruments networks. 3rd WMO Scientific Conf. on Weather Modification (Vol. II), Clermont-Ferrand, France.

Musil, D.J., A.J. Heymsfield, and P.L. Smith, 1986: Microphysical characteristics of a well-developed weak echo region in a High Plains supercell thunderstorm. *J. Appl. Meteor.*, **25,** 1037–1051.

Orville, H.D., 1996: A review of cloud modeling in weather modification. *Bull. Amer. Meteor. Soc.*, **77,** 1535–1555.

Orville, R.E., 1994: Cloud-to-ground lightning flash characteristics in the contiguous United States 1989–91. *J. Geophys. Res.*, **99,** 10833–10841.

Pasquill, F., 1974: *Atmospheric diffusion*, 2nd ed. John Wiley & Sons, New York.

Pham Van Dinh, 1973: Mesure du rendement des générateurs de particules AgI-NaI avec différentes méthodes d'échantillonnage. *Anelfa*, **21,** 30–36.

Prodi, F., 1974: Climatologia della grandine nella Valle Padana (1968–1971). Climatology of hail in the Po Valley (1968–1972). *Revista Italiana di Geofisica*, **23** (516), 283–290.

Reinking R.F., S.Y. Matrosov, R.A. Kropfli, B.E. Martner, and B.W. Bartram, 1998: Identification of freezing drizzle and ice crystal types with dual-polarization K_a–band radar. Proceedings of 13th AMS Conf. on Planned and Inadvertent Weather Modification, Everett, Washington, August 17–21, 1998, 249–252.

Rilling, R.R., R.R. Blumenstein, W.G. Finnegan, and L.O. Grant, 1984: Characterization of silver iodide-potassium iodide ice nuclei: rates and mechanisms, and comparison to the silver iodide-sodium iodide system. 9th WMO Conf. on Weather Modification, Park City, Utah, 16–17.

Sánchez J. L., R. Fraile, and A. Vega, 1998: Microphysical aspects of a hailstorm seeded with ground generators network in Lleida (Spain). Proceedings of 13th AMS Conf. on Planned and Inadvertent Weather Modification, Everett, Washington, August 17–21, 1998, J44-J47.

Sánchez J.L., R. Fraile, J.L. de la Madrid, M.T. de la Fuente, P. Rodriquez, and A. Castro, 1996: Crop damage: the hail size factor. *J. Appl. Met.,* **35,** 1535–1541.

Sánchez J.L., J.L. Marcos, M.T. de la Fuente, and R. Fraile, 1994: Criteria for a remote ground generator network. *J. Weather Modif.,* **26,** 83–88.

Schaefer, V.J., 1946: The production of ice crystals in a cloud of supercooled water droplets. *Science,* **104,** 457.

Schleusener, R.A., and P.C. Jennings, 1960: An energy method for relative estimates of hail intensity. *Bull. Amer. Meteor. Soc.,* **41,** 372–376.

Smith, C., 1999: Personal communication.

Smith, P.L., L.R. Johnson, D.L. Priegnitz, B.A. Boe, and P.J. Mielke, Jr., 1997: An exploratory analysis of crop hail insurance data for evidence of cloud seeding effects in North Dakota. *J. Appl. Meteor.,* **36,** 463–473.

Stone, R.E., 1998: Use of a new targeting model in the design of winter cloud seeding programs. Preprints, 13th AMS Conf. on Planned and Inadvertent Weather Modification, Everett, Washington, August 17–21, 1998, 626–629.

Vonnegut, B., 1947: The nucleation of ice formation by seeding with dry ice. *J. Appl. Phys.,* **18,** 593–595.

Vonnegut, B., 1981: Misconceptions about cloud seeding with dry ice. *J. Weather Modif.,* **13,** 9–10.

Warburton, J.A., G.O. Linkletter, and R. Stone, 1982: The use of trace chemistry to estimate seeding effects in the national hail research experiment. *J. Appl. Meteor.,* **21,** 1089–1110.

WMA, 1986: Weather Modification Association weather modification capability statement. *J. Weather Modif.,* **18,** 141–142.

WMA, 2000: Qualifications and procedures for certification. *J. Weather Modif.,* **32,** 93–94.

WMO, 1992: Statement on the status of weather modification. *J. Weather Modif.,* **25,** 1–6.

WMO, 1996: Meeting of experts to review the present status of hail suppression. Programme on Physics and Chemistry of Clouds and Weather Modification Research, Golden Gate Highlands National Park, South Africa, November 6–10, 1995. WMO Technical Document No. 764, WMP Report No. 26, 39 p.

WMO, 1997: Register of national weather modification projects 1996. Programme on Physics and Chemistry of Clouds and Weather Modification Research. WMO Technical Document No. 939, WMP Report No. 32, 49 p.

8.0 CONVERSION OF UNITS

Manuals of Standard Practice specify units in accordance with the International System of Units, or SI (from the French, *Le Système International d'Unités*). In the SI, the base unit of length is the meter (m); of mass, the kilogram (kg); of time, the second (s); and of temperature, the Kelvin (K). Derived SI units include those for area, the square meter (m^2); for volume, the cubic meter (m^3); and for speed, a meter per second ($m\ s^{-1}$). Other derived SI units relevant to this manual include force, the newton (N, or $m \cdot kg \cdot s^{-2}$); pressure, the pascal (Pa, or $N \cdot m^{-2}$); and energy, the joule (J, or $N \cdot m$). Also included in the derived SI units is Celsius temperature (°C), which is equivalent to $K - 273.16$.

In addition, a number of prefixes are approved for use with SI units, so that very small or very large quantities need not be rendered in unwieldy formats. Such prefixes are listed in Table 8-1, below.

These prefixes are approved for use with SI units, so that 1,000 m also can be expressed as 1 km. Likewise, one-millionth of a meter can be expressed as 1 μm. In this document, these approved SI prefixes are used wherever applicable to reflect the common usage. Examples are listed in Table 8-2. Alternate common English equivalents, other common units, and the conversion factors also are shown. A limited number of other units outside the SI are approved for use with the SI. Among these are minutes (min), hours (h), days (d), and liters (L).

TABLE 8-1. Approved SI Prefixes

Factor	Name	Symbol
10^6	mega	M
10^3	kilo	k
10^2	hecto	h
10^{-2}	centi	c
10^{-3}	milli	m
10^{-6}	micro	μ

TABLE 8-2. Conversion Factors

Quantity	SI Units	Equivalent cgs Units	Other Common Units	Equivalent English Units
length	1 meter (m)	100 centimeters (cm)	3.28 feet (ft)	1.09 yards (yd)
	1000 m		1.0 kilometer (km)	0.54 nautical mi
	1000 m		1.0 km	0.62 statute mi
	0.001 m		1.0 millimeter (mm)	0.039 inches (in.)
area	1 square meter (m^2)	10,000 cm^2	10.8 ft^2	1.2 yd^2
	1,000,000 m^2		1.0 km^2	0.39 mi^2
volume	1 cubic meter (m^3)	1,000,000 cm^3	35.3 ft^3	1.31 yd^3
	1,000,000,000 m^3		1.0 km^3	0.24 mi^3
	10^{-6} m^3	1 cm^3	1 cc	0.034 fluid ounces (fl oz)
mass	1 kilogram (kg)	1000 grams (g)	35.36 ounces (oz)	2.21 pounds (lb)
pressure	1 Pascal (Pa)	10 dynes cm^{-2}	0.098 millibar (mb)	0.0003 in. of Hg
speed	1 meter per second (m s^{-1})	100 cm s^{-1}	3.6 km h^{-1}	2.24 mi h^{-1}
temperature	273.16 Kelvin (K)		0.0 degrees Celsius (°C)	32.0 degrees Fahrenheit (°F)
			°C = K − 273.16	°F = (1.8°C) + 32
			K = °C + 273.16	°C = 5/9 (°F − 32)
time	1 second (s)	1 s		1 s
	60 s		1 minute (min)	1 minute (min)
	3,600 s		1 hour (h)	1 hour (h)
	86,400 s		1 day (d)	1 day (d)

INDEX

(f), page with figure; (t), page with table